International Workshop on Transportation and Supply Chain Engineering (IWTSCE'16)

MATEC Web of Conferences Volume 105 (2017)

Rabat, Morocco
29 - 30 November 2016

Editors:

B. Abou El Madj
M. Bouya

ISBN: 978-1-5108-3939-7

Printed from e-media with permission by:

Curran Associates, Inc.
57 Morehouse Lane
Red Hook, NY 12571

Some format issues inherent in the e-media version may also appear in this print version.

Printed by Curran Associates, Inc. (2017)

For additional information, please contact EDP Sciences – Web of Conferences
at the address below.

EDP Sciences – Web of Conferences
17, Avenue du Hoggar
Parc d'Activité de Courtabœuf
BP 112
F-91944 Les Ulis Cedex A
France

Phone: +33 (0) 1 69 18 75 75
Fax: +33 (0) 1 69 28 84 91

contact-edps@webofconferences.org

Additional copies of this publication are available from:

Curran Associates, Inc.
57 Morehouse Lane
Red Hook, NY 12571 USA
Phone: 845-758-0400
Fax: 845-758-2633
Email: curran@proceedings.com
Web: www.proceedings.com

TABLE OF CONTENTS

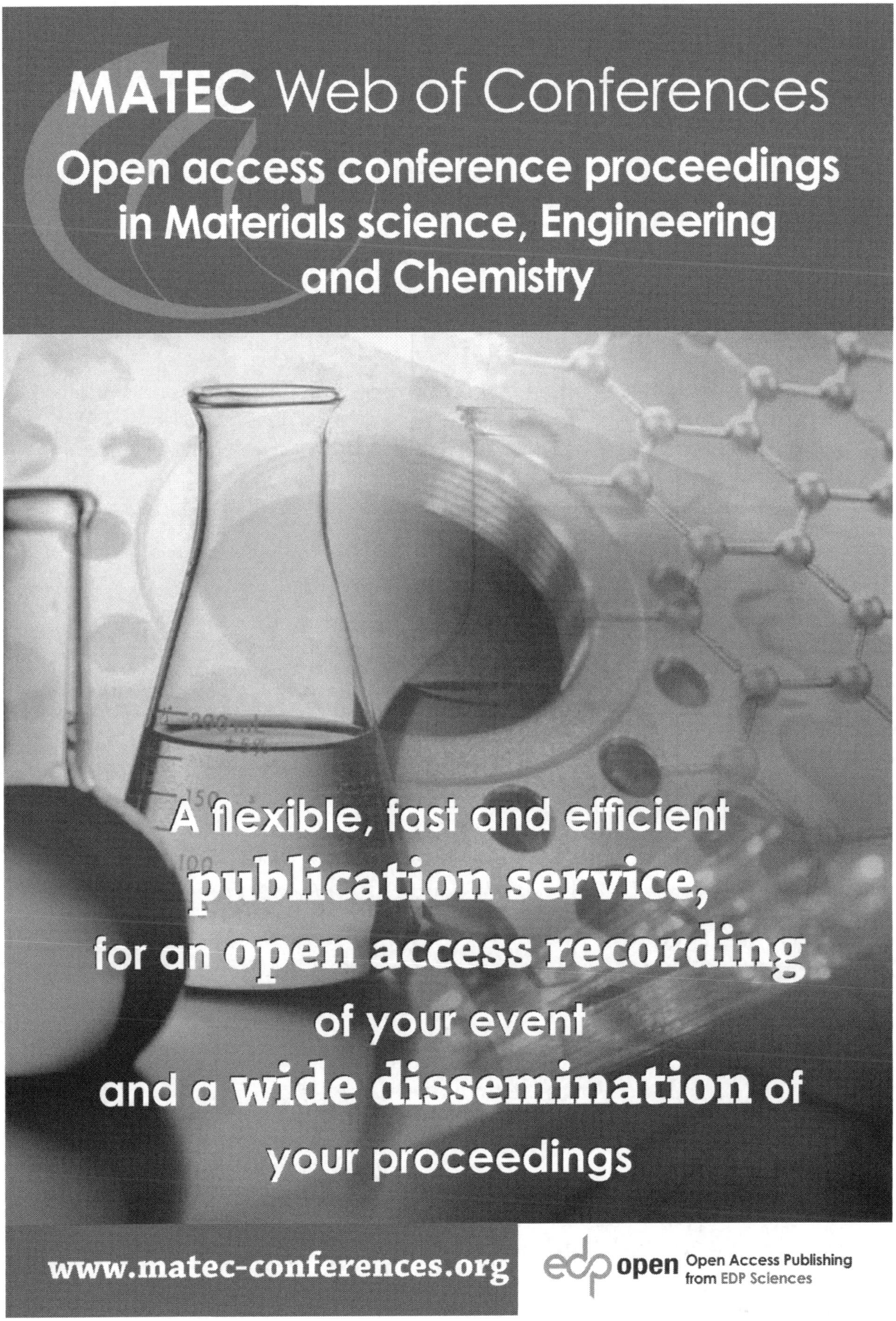
MATEC Web of Conferences
Open access conference proceedings in Materials science, Engineering and Chemistry
A flexible, fast and efficient
publication service,
for an open access recording
of your event
and a wide dissemination of
your proceedings
www.matec-conferences.org
edp open
Open Access Publishing from EDP Sciences

Preface:

International Workshop on Transportation and Supply Chain Engineering (IWTSCE'16) is organized by Hassan II University of Casablanca and International University of Rabat. The purpose of this meeting is to provide an international forum for recent developments in selected fields of research in transportation systems and supply chain engineering by bringing together researchers, practitioners and research scholars from all over the world to exchange and share their experiences and research results. The workshop is intended to provide chances for intensive discussions about their own selected topics, practical challenges encountered and identify future directions for the development of related modeling strategies and computational techniques. Innovative presentations are all invited, and participants for auditing only are also welcomed.

In order to make IWTSCE conference proceedings widely usable by the scientific community, we are proud having a partnership with MATEC as one of reputable partner with highly flexible publishing solutions. All accepted paper will be published by EDP Sciences on EI Serial Journal of MATEC web of Conferences {ISSN (Electronic Edition):2261-236X} as a proceeding. MATEC Web of Conferences is indexed in SCOPUS, Thomson Reuters (web of science), Chemical Abstracts Service (CAS), DOAJ, Engineering Village, Google Scholar, and Inspec. For further information about IWTSCE, please visit the web site at: www.iwtsce16.eb2a.com.

We express our deepest gratitude to all members of the scientific committee, invited speakers, reviewers, sponsors, participants in the conference and members of the IWTSCE organizing committee for their valuable contribution to the successful organization of IWTSCE'16.

Dr. Badr Abou El Majd

General Chair
International Workshop on Transportation and Supply Chain Engineering

edp sciences

MATEC Web of Conferences

Statement of Peer review

In submitting conference proceedings to MATEC Web of Conferences, I certify to the Publisher that I adhere to the **Policy on Publishing Integrity** of the journal in order to safeguard good scientific practice in publishing.

1. All articles have been subjected to peer review administered by the proceedings editors.
2. Reviews have been conducted by expert referees, who have been requested to provide unbiased and constructive comments aimed, whenever possible, at improving the work.
3. Proceedings editors have taken all reasonable steps to ensure the quality of the materials they publish and their decision to accept or reject a paper for publication has been based only on the merits of the work and the relevance to the journal.

Title, date and place of the conference

International workshop in Transportation and Supply Chain Engineering (IWTSCE'16)
November 29 - 30, 2016
International University of Rabat
Rabat, Morocco

Proceedings editor(s):

Editor: B. Abou El Majd, University Hassan II of Casablanca
Associate Editor: M. Bouya, International University of Rabat

Date and editor's signature

22 March 22, 2017

MATEC Web of Conferences 105, 00001 (2017) DOI: 10.1051/matecconf/201710500001
IWTSCE'16

Solving stochastic multiobjective vehicle routing problem using probabilistic metaheuristic

Asmae Gannouni[1,2,⋆], *Rachid* Ellaia[1], and *El-Ghazali* Talbi[2]

[1] *LERMA, Mohammadia School of Engineering, Mohammed V University Rabat, Morocco*
[2] *CRIStAL/Dolphin, University Lille 1*

Abstract. In this paper we propose a new metaheuristic algorithm for solving stochastic multiobjective combinatorial optimization(SMOCO) problems. Indeed, we find that the various initiatives that have been launched recently on this subject, they propose the classical metaheuristics to solve a stochastic multi-objective problems, but when the stochasticity effects is not taken into account, the choice of an arbitrary value leading to a particular configuration and a high loss of information. To conserve the stochastic nature of SMOCO problems, the pareto ranking should be defined on the random objectives functions directly rather than converted deterministic objectives. From these considerations, the scope of this research should consist of: (i)Proposed novel methodology of stochastic optimality for ranking objective functions characterized by non-continuous and no closed form expression. This novel approach is based on combinatorial probability and can be incorporated in a multiobjective evolutionary algorithm. (ii)Provide probabilistic approaches to elitism and diversification in multiobjective evolutionary algorithms. Finally, The behavior of the resulting Probabilistic Multi-objective Evolutionary Algorithms (PrMOEAs) is empirically investigated on the multi-objective stochastic VRP problem.

1 Introduction

Many real-life optimization problems encountered in logistics, transportation, and Markets finance have several conflicts objectives under aleatory uncertainty to be satisfied, traditionally such problems were handled by converting the stochastic multiple objectives to deterministic multiple objectives using statistical aggregation: mean value, extreme value, variance..., then applying the classical metaheuristics [1], such as pareto evolutionary algorithms for generate a set of well-distributed pareto-optimal solutions [13]. Such an approach has many problems, the including the loss of significant trade-off information and the inability to search the true objective space because it is incapable to describe the relationship among stochastic(random) objectives. Personally speaking, to represent and conserve the stochasticity of original SMOCO problems, we should proposed other methodology of optimality under aleatory uncertainty. Therefore, the aim of this study is to provide a probabilistic definition, to dominance, elitism and diversification in multi-objective evolutionary algorithms.

Based on the literature, the authors have classified methods for solving stochastic multiobjective combinatorial optimization problems, into three specific categories based on the way the objective function is estimated: i)Deterministic approaches: have been focused on treating it in multiobjective design by avoiding or ignoring uncertainty[18], then the solution of the stochastic multiobjective optimization problems is not essentially different from that of a deterministic multiobjective optimization problems, and exact(lerning method, random optimization methods, stochastic approximation) [5] or metaheuristic techniques(NSGAII, SPEAII,...) can be used [4,15] easily. ii) Robust Approaches: In this categorization, researchers reduced the stochastic nature of the problem by using for example the expectation of random functions: sample average approximation [17], or pareto means, pareto variance, pareto Quantile,...[6,12], these approaches are similar to the deterministic approaches, pareto dominance is used after changing the methodologies of the evaluation. Finally, for (iii)Stochastic Approaches: only few techniques are available and are still in nascent [11,16], this work is part of these Approaches, here the stochastic optimality is the major challenge. These approaches are well the resulting of hybridization between the probabilistic programming, simulation [8, 10] and metaheuristic algorithms.

The remainder of this paper is organized as follows: In section 2, we investigate the formula of stochastic multi-objective optimization problems and we discuss the main concepts in multiobjective evolutionary algorithms that can be extended in stochastic context. Section 3, describes the design of Probabilistic approaches: probabilistic dominance, probability memory and diversification for practical use. He whole methodology is then illustrated on analytical test case: Multiobjective stochastic VRP in section

⋆e-mail: asmaegannouni@gmail.com

4, followed by conclusions and future prospects in Section 5.

2 Stochastic Multi-objective Optimization

Stochastic multi-objective optimization problems[7] have been investigated thoroughly in the operations research community. We start this section by giving a general formula of SMOCO problems, we then expose general concepts that share these problems and the challenge that it entails. Then we present pareto multi-objective evolutionary algorithms which will be be extended on stochastic context[21-23] in the next section.

2.1 Stochastic Multi-objective Optimization Problems

In the stochastic multi-objective optimization problems, two or more usually conflicting stochastic(uncertain) objectives are required to be optimized simultaneously, several types of formulations for SMOCO problems are possible, to be more specific let us define the problem in terms of the minimization that motivated the present research:

$$\begin{array}{c} \min_{X\times\Omega}\ F(x,\omega) = (f_1(x,\omega), ..., f_n(x,\omega)) \\ subject\ to: g_i(x,\omega) \leq 0,\ i = 0, ..., m \end{array} \quad (1)$$

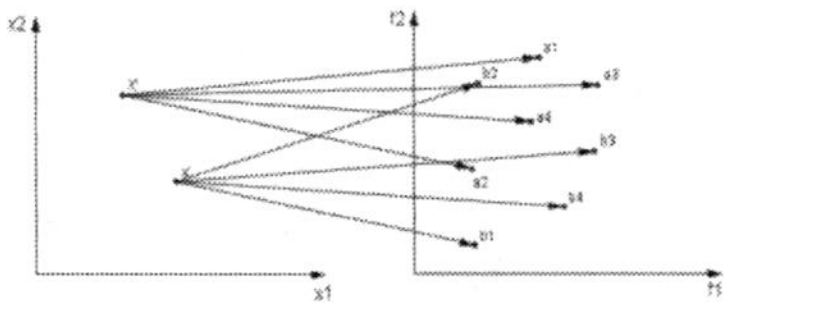

Figure 1. Evaluations of two individuals x and x'

Where $f_i(x,\omega)$ $(i = 1, ..., n)$(for each fixed x is the set of decision variables from the decision space X), are Borel functions(random variables) defined on a probability space $(\Omega, \mathcal{F}, \mathbb{P})$ and ω is an element of the sample space Ω. Each ω determines a specific realization (a scenario)of the random variable $f_i(x, .)$.(x it is called an individual in evolutionary multiobjective optimization algorithms MOEAs). The probability space $(\Omega, \mathcal{F}, \mathbb{P})$ is defined as follows:
- Ω is the space of elementary events.
- $\mathcal{F}$ is the σ-algebra constructed from all subsets of Ω.
- $\mathbb{P}$ is a probability measure defined on the measurable space $(\Omega, \mathcal{F})$.

From this formulation of SMOCO problems, each objective function can yield different values for the same individual from time to time (Figure.1), concerning the multi-objective part, we focus on the solution concept of determining pareto-optimal solutions, under this approach there are three possible relationships between the exact solutions:
Definition.1 (Weak Dominance) x weakly dominates x' denoted by $x \preceq x'$ if $f_i(x) \leq f_i(x'), \forall i \in \{1, ..., m\}$.
Definition.2 (Strong Dominance) x strongly dominates x' denoted by $x \preceq x'$ if $f_i(x) < f_i(x'), \forall i \in \{1, ..., m\}$.
Definition.3 (Incomparable) x is incomparable with x' denoted by $x \sim x'$ if $f_i(x) > f_i(x'), \exists i \in \{1, ..., m\}$ and $f_j(x) > f_j(x') \exists j \in \{1, ..., m\}$.
Definition.4 (Nondominated vector) An objective vector $F(x)$ is said to be nondominated iff there does not exist another objective vector $F(x')$ such that $x' \succ x$.
Definition.5 (Efficient solution) A solution $x \in X$is said to be efficient(or pareto optimal, nondominated) iff its mapping in the objective space results in a nondominated vector.
The set of all efficient solutions is called efficient(or Pareto optimal) set, and its mapping in the objective space is called Pareto front. A possible approach in multiobjective combinatorial optimization problem solving is to find the minimal set of efficient solutions. But, generating the entire set of pareto optimal solutions is usually, due to the complexity of the underlying problem or to the large number of optima. Therefore, the overall goal is often to identify a good efficient set approximation in stochastic objectives space. These relationships can not compared objec-

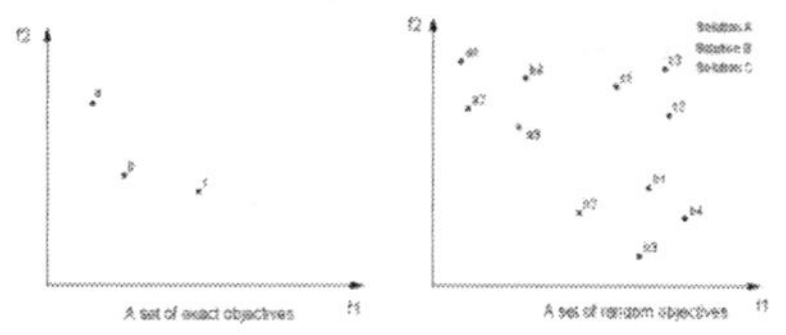

Figure 2. Difference between exact and random solutions

tives value samples $\{a_i\}_{i=1}^4$ and $\{b_i\}_{j=1}^4$ (Figure2) of given two individuals x and x', from this limitation the purpose of this study is to determine stochastic order to describe the relationship among random objectives.

2.2 Multiobjective Evolutionary Algorithms

Multiobjective evolutionary algorithms(MOEAs) is a class of stochastic search techniques within metaheuristics that have already been successfully applied to a wide range of optimization problems under uncertainty [21,22]. Research in the design of multi-objective evolutionary algorithms has mainly focused on the fitness assignment, elitism and diversification. We present briefly the notion of these three components:
- Fitness assignment: A notion of Pareto dominance plays an important role to seek Pareto optimality in EMOAs the most example of this component is dominance rank approach.
- Elitism: Is implemented by selecting the best individuals to a mating pool through a such as binary tournament selection of the combined archive and evolving population.
- Diversification: The aim of Diversity maintenance in objective space is to enable a good approximation of the whole Pareto set. An example of diversity strategies are,

nearest neighbour method and crowding distance.
Pareto multi-objective evolutionary algorithms (PMOEAs) operate on these common principles:

Algorithm 1 MOEAs

1: Initialization: Generate random initial population P.
2: Evaluation: Evaluate individuals in P according to Pareto dominance ralation.
3: Elitism: Select elite solutions from P to creat or update archive.
4: Variation: Apply crossover and mutation operator.
5: Stopping condition: if it is satisied return A, else go to step 2.

The previous concepts are especially for a deterministic multiobjective case, where the solutions are exact values, we propose in the next section to extended them in stochastic context, by using combinatorial probability.

3 Probabilistic Approaches in Multi-objective Evolutionary algorithms

A hybridization of probabilistic programming with metaheuristics have attracted a growing interest during the last few years. This work is part of this research community, the importance of the hybridization proposed in this study is use the probabilistic programming for handling aleatory uncertainty in evolutionary algorithms[24], we choose a conceptually simple approach that is also frequently applied to stochastic programming, namely the optimization based on a fixed sample of random scenarios(i.e, the scenario model is applicable only when the underlying probability distribution is discrete). In this approach, the given distribution of the random vector ω is approximated by an empirical sample distribution(discrete uniform,binomial ...), obtained by drawing N scenarios $\{\omega_l\}$ $(l = 1, ..., N)$, from the original distribution of ω, and replacing the original distribution(continuous or discrete) by a discrete distribution(i.e, A discretization of exponential distribution gives us the binomial distribution) where each of the N scenarios has the same probability 1/N(1/N in equi-probable sampling). After evaluation technique, the challenging is to estimate the impacts of uncertainty on the samples through a non-parametric statistics: Stochastic order, more specifically probabilistic dominance, then this mathematical instrument is introduced in MOEAs to describe the relationship among random objectives and exploration of the stochastic solutions space. Many stochastic orders [2] exist and can be introduced in MOEAs,the appropriate choice between these relations, is depending on the problem being studied. Briefly, this section is devoted to the application of probabilistic dominance to the with stochastic multi-objective combinatorial programming problems. This way of ranking is based on combinatorial probability. Then the formal definition of elitism and diversification in stochastic context are proposed.

3.1 The proposed methodology of Probabilistic Optimality

In the stochastic case, the objective values are different each time a individual is evaluated. So, the vector function F does not represent a deterministic mapping from the decision space to solution space (Figure.1), because an infinite set of different objective vectors is now assigned to a solution $x \in X$.

In the present work, we choose a conceptually simple approach for probabilistic dominance that compare two stochastic(random) objective vectors.

Definition.6 Given two individual x and x', the probability that solution $F(x, \omega)$ a dominates solution $F(x', \omega)$ is given by:

$$x \prec_{pr} x' \quad \Leftrightarrow \quad Pr(F(x,\omega) \prec F(x',\omega)) > Pr(F(x',\omega) \prec F(x,\omega)) \tag{2}$$

where $Pr(F(x,\omega) \prec F(x',\omega)) =$

$$\sum_{i=1}^{l_1} \sum_{j=1}^{l_2} 1/l_1 . 1/l_2 . I(F(x,\omega_i) \prec F(x',\omega_j))$$

is the non-parametric order relation,and note that (2) is the well known Wilcoxon-Man-Whitney statistic. Which is a non-parametric (i.e., distribution-free) statistical, it takes objective value samples of given two individuals x and x', and this probabilistic order which reflects the stochastic nature of the problem in the results.For a perfect ranking function the $Pr(F(x,\omega) \prec F(x',\omega))$ statistic is 1 and 0 in Incomparable case.

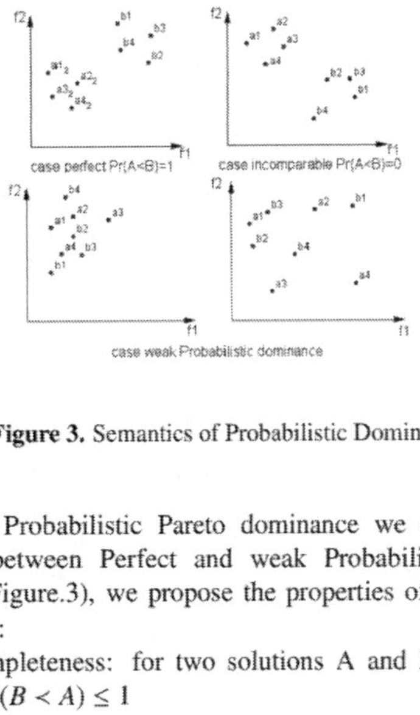

Figure 3. Semantics of Probabilistic Dominance

Within Probabilistic Pareto dominance we can distinguish between Perfect and weak Probabilistic dominance(Figure.3), we propose the properties of this order relation:
- Incompleteness: for two solutions A and B, $Pr(A \prec B) + Pr(B \prec A) \leq 1$
- Weak Probabilistic transitivity: let us consider three solutions A, B and C:

1. if $Pr(A \prec B) \leq 0.5$ and $Pr(B \prec C) \leq 0.5$ then $Pr(A \prec C) \leq 0.5$.

2. if $Pr(A \prec B) = 1$ and $Pr(B \prec C) = 1$ then $Pr(A \prec C) = 1$.

3.2 Fitness assignment scheme

The main role of this procedure is to guide the search algorithm toward Pareto optimal solutions for a better convergence, but the probabilistic order relation cannot be used to define the layers on solution space, then what is the Rank of each solutions in S?

Definition.7 The rank of each solution $A = \{a_1, a_2, ..., a_l\} \in S$ is determined by the probabilities of other solutions $O = \{o_1, o_2, ..., o_l\} \in S$ dominating it :

$$Rank(A) = Pr(a_i \in A) \times \prod_{O \neq A}(1 - Pr(O < a_i) \quad (3)$$

$$Rank(A) = \frac{1}{l}\sum_{i=1}^{l}\prod_{A \neq O}(1 - \frac{|\{o \in O \setminus o < a_i\}|}{|O|})) \quad (4)$$

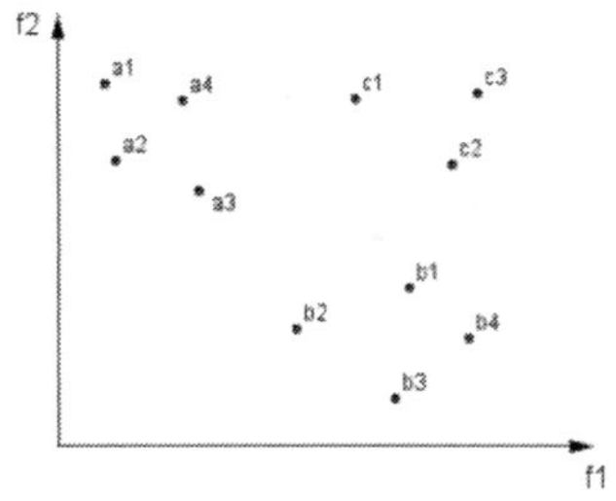

Figure 4. A set of random objectives

$$Rank(a_i) = (1 - \frac{|\{o \in O \setminus o < a_i\}|}{|Y|}) \quad (5)$$

Example 1: We consider a set of solutions $S = \{A, B, C, D, ..\}$ (Figure.4).
a- $|A| = 4$, $|C| = 3$, c_1 , c_2 and c_3 are dominated by a_3, $Pr(A < C) = 0.5 \Rightarrow$ Solution A dominates weakly C.
a- Solution A is non-dominated by any solution $\in S \Rightarrow Rank(A) = 1$, similarly $Rank(B) = 1$.

3.3 Elitism:Probability Memory

Elitism has always been a major topic of discussion in the evolutionary computation community. In stochastic context, the question of when to accept a newly generated solution despite those previously found, is palliated by the probabilistic memory. The decision to add any solution to the archive is based on a mean probability that the solution is dominated by other solutions in S.

Definition.8 A Probabilistic archiving schema is defined with this relation:

$$PrArchiv(p) = \{A \in S \setminus Pr(A) \geq p\} \quad (6)$$

where Pr(A)=Rank(A) and $p \in [0, 1]$ is the fixed probability(chosen by a decision maker) for archiving the best solutions.

3.4 Diversification

In stochastic case, diversification is preserved in the re-sampling procedure. The probability that two solutions are equal is a rare event(for testing A=B uses Jaccard similarity coefficient). A few work that is proposed the modification of this technique, for example the crowding distance is a measure of how close an individual is to its neighbors ; large average crowding distance will result in better diversity in the population.We can Define a probabilistic crowding distance [18] as follows:

$$PC_i = \sum_{k=1}^{m} \frac{\mu(f^k_{i+1}) - \mu(f^k_{i-1})}{\mu(f^k_{max}) - \mu(f^k_{min})}$$

3.5 The result of Hybridization:PrMOEA

The main steps of the Probabilistic Multi-objective Evolutionary Algorithms(PrMOEAs) within an iteration are shown below:

1. Evaluate all solutions in initial population.
2. Fitness assignment: Calculate the probabilistic Pareto ranking of all solutions.
3. Elitism Maintain the best p% of the solutions in terms of fitness and discriminate the rest.
4. Perform selection, crossover, and mutation on the population.

We investigate the performance of this algorithm for solving the stochastic version of multi-objective vehicle routing problem.

4 Case study:Stochastic Multi-objective VRP

Vehicle routing problem have received increasing attention in recent years. However, in practical situations, the deterministic and single VRP cannot cover real-life contexts, the objectives can be multiple [20] and some components of the VRP are random [25]. In this work, we are interested an important variant of VRP, the so called multiobjective stochastic VRP with two stochastic objectives. A classification of VRP according to the stochastic parameters is proposed in [24, 25], and one of the most commonly applied approaches to SVRPs is stochastic programming (SP) where a general introduction is given in [3].

4.1 Mathematical Formulation

Let G(V,E) be a directed graph, where E is the set of arcs and V is the set of nodes: $V = \{0, 1, ..., n-1, n\}$ represents the n customers to be visited. A feasible route is represented by the set of served customers, starting and ending at the central depot: $R = \{0, 1, 2...n-1, n, 0\}$ for simplicity, the depot is denoted as customer 0, the depot is treated as the source of service demanded by the customers. The fleet of vehicles, indexed by $k = \{1, 2, .., K\}$ is based at the depot, and all the routes must start and end at the depot. The objective of this problems is to find optimal routes with respect to the following two stochastic objectives functions:

$$f_1(x, \omega) = \sum_{i=0}^{n} \sum_{j=0 j \neq i}^{n} \tilde{T}_{ij} x_{ij}^k \quad (7)$$

$$f_2(x, \omega) = \sum_{i=0}^{n} \sum_{j=0 j \neq i}^{n} \tilde{D}_{ij} x_{ij}^k \quad (8)$$

Where $\tilde{T}_{ij}$ is a aleatory duration of travelling from i to j, the traveling time between two customers i and j and $\tilde{D}_{ij}$ is a aleatory travelled distance. These two component are stochastic variables depend on traffic, weather or road conditions or customer requirements such as stochastic demand. The main constraints of MSVRP-SD are the following:

- **Vehicle capacity constraint**: imposes that the total of stochastic customers' demands for any route must not exceed the limited vehicle capacity.
- **Distance constraint**: imposes that each vehicle with a limited capacity must deliver goods to the customers with the minimum transportation costs in term of travelled distance. Yet, if the capacity constraint of a vehicle is not satisfied, the delivery fails and causes wasted costs.

In order to resolve and find good solutions for this problem, we propose two probabilistic metaheuristic algorithms: PrMOEAs and PrNSGAII.

4.2 Solution techniques:PrMOEAs

To solve MSVRP-SD, we propose a Modified NSGAII. Other contribution in this paper, the NSGAII [9] is adapted to stochastic context by integrating the proposed probabilistic approaches defined in the previous section. PrMOEAs was developed under C++ framework, and PrNSGAII is implemented by modiefied NSGAII under ParadisEO-MOEO [19]. We test a PrNSGAII algorithm on a number of solomon's benchmarks instances: C101, C201, R101, R201 modified. with time windows and stochastic service times and stochastic demands.

Table 1. Parameter Configurations for PrMOEAs

Parameter	Value
Population size	100
Number of Samples	25
Max of generation	500
Crossover rate	0.9
Mutation rate	0.2
Probability p	0.7

Subsequently, we have obtained encouraging results (Figure.6). To evaluate the performance of the algorithm we can compute the means of every stochastic solution and compared it with a solution in deterministic case (Figure.5).

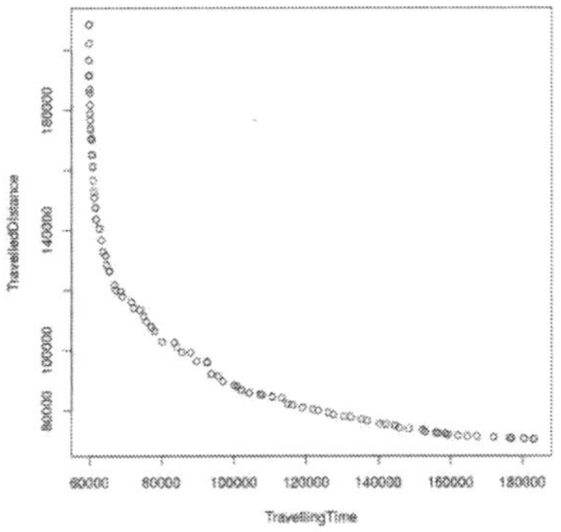

Figure 5. Deterministic Pareto Front(C101-NSGAII)

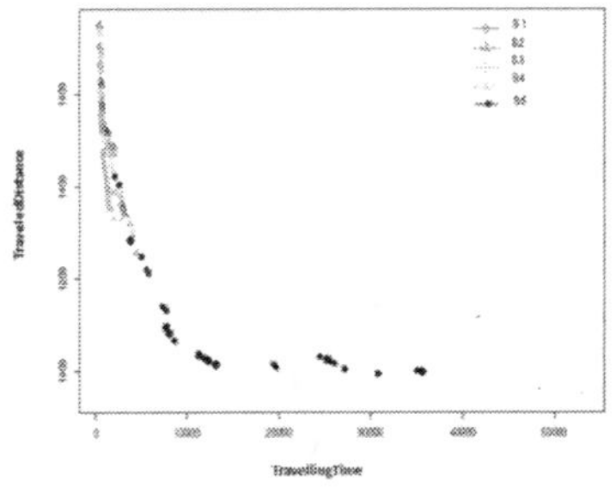

Figure 6. Probabilistic Pareto Front(C101-PrNSGAII)

5 Conclusion and Discussion

In this work, an algorithm for stochastic multiobjective combinatorial optimization(SMOCO) problems has been proposed and analyzed. The algorithm is obtained by using a hybridization of probabilistic programming with metaheuristic. Among the main objectives of this hybridization: we extended the dominance criteria, elitism and diversification by using combinatorial probability and non-parametric statistics. The validity of proposed algorithm is empirically investigated and the results demonstrate the capabilitie of this technique to generate a set of well-distributed probabilistic pareto-optimal solutions As future work, it will be also interesting to develop Performance Evaluation, we plan the inclusion of the robustness criteria with Bayesian probability and sensitivity analysis, which aims to judge the quality of stochastic solutions. For testing these novel approaches, the applications can be found in the logistic field, such as Green Transportation problems.

References

[1] El Ghazali Talbi, *Metaheuristics:From Design to Implementation* (Wiley, 2009) 624

[2] Moshe Shaked, J. George Shanthikumar, *Stochastic Orders* (Springer,New York, 2007) 473

[3] John R. Birge, François Louveaux, *Introduction to stochastic programming* (Springer, New York, 2011) 500

[4] Walter J. Gutjahr, *Two metaheuristics for multiobjective stochastic combinatorial optimization*. Stochastic Algorithms:Foundations and Applications. **3777**, Lecture Notes in Computer Science, 116-125, (2005)

[5] Norio Baba, Akira Moromito. *Three approaches for solving the stochastic multi-objective programming problem*. Stochastic Optimization. Lecture Notes in Economics and Mathematical Systems, **379**, (Springer, Berlin, Heidelberg),93-109 (1992)

[6] Xinchang Hao, Mitsuo Gen, Lin Lin, Gursel A.Suer. *Effective multiobjective EDA for bi-criteria stochastic job-shop scheduling problem*. Journal of Intelligent Manufacturing, 1-13, (2015)

[7] Giovanni Petrone, Gianluca Iaccarino, D. Quagliarella. *Robustness Criteria in optimisation under uncertainty*. Evolutionary and deterministic methods for design, Optimisation and control. CIRA, Capua, Italy 2011

[8] Walter J. Gutjahr, Alois Pichler, *Stochastic multi-objective optimization: a survey on non-scalarizing methods*, Annals of Operations Research **236**, Issue 2, pp 475–499, (January 2016)

[9] Angel A.Juan, Javier Faulin, Scott E.Grasman, Markuts Rabe, Gonçalo Figueira. *A review of simheuristics: Extending metaheuristics to deal with stochastic combinatorial optimization problems*. Operations Research Perspectives **2** , 62-72, (2015)

[10] Kalyanmoy Deb, Samir Agrawal, Amrit Pratap, T Meyarivan, *A fast elitist non-dominated sorting genetic algorithm for multi-objective optimization: NSGA-II*, Parallel Problem Solving from Nature PPSN VI. Lecture Notes in Computer Science **1917**, pp 849-858, (2000)

[11] Josef Schwarz and jiri Ocenasek, *Multiobjective bayesian optimization algorithm for combinatorial problems:Theory and practice*, Neural Network World **11**(5), 423-441, 2001

[12] Yang Liu, Zhi-Ping Fan, Yao Zhang, *A method for stochastic multiple criteria decision making based on dominance degrees*. Information Sciences **181**, 4139-4153, (2011)

[13] Rajan Filomeno Coelho. *Probabilistic Dominance in Multiobjective Reliability-Based Optimization: Theory and Implementation*. IEEE Transactions on evolutionary computation **19**, No. 2, 214-224, (2015)

[14] Henri Bonnel, Julien Collonge. *Optimization over the Pareto outcome set associated with a convex bi-objective optimization problem: theoretical results, deterministic algorithm and application to the stochastic case*, Journal of Global Optimization **62**, 481–505, (2015)

[15] Henri Bonnel, Julien Collonge.*Stochastic Optimization over a Pareto Set Associated with a Stochastic Multi-Objective Optimization Problem*. Journal of Global Optimization **162**, 405–427, (2014)

[16] Mathieu Basseur, Eckart Zitzler. *Handling Uncertainty in Indicator-Based Multiobjective Optimization*. International Journal of Computational Intelligence Research **2**(3), (2006)

[17] Jörg Fliege, Huifu Xu. *Stochastic Multiobjective Optimization: Sample Average Approximation and Applications*. Journal of Optimization Theory and Applications **151**, Issue 1, pp 135–162, (2011)

[18] Rafael Caballero, Emilio Cerdá, Maria M. Munoz, Lourdes Rey. *Stochastic approach versus multiobjective approach for obtaining efficient solutions in stochastic multi-objective programming problems*. European Journal of Operational Research **158**, 633–648, (2004)

[19] Arnaud Liefooghe, Matthieu Basseur, Laetitia Jourdan, El-Ghazali Talbi. *ParadisEO-MOEO: A Framework for Evolutionary Multi-objective Optimization*. Evolutionary Multi-Criterion Optimization, Springer, pp 386-400, (2007)

[20] Nicolas Jozefowiez, Frédéric Semet, El-Ghazali Talbi. *A Multiobjective vehicle routing problems*. European Journal of Operational Research **189**,pp 293-309, (2008)

[21] Evan Hughes. *Evolutionary multi-objective ranking with uncertainty and noise*. In Proceedings of the First Conference on Evolutionary Multi-Criterion Optimization,pp 329–343, (2001)

[22] Kay Chen Tan, Chi Keong Goh. *Handling Uncertainties in Evolutionary Multi-Objective Optimization*. Computational Intelligence: Research Frontiers. Lecture Notes in Computer Science **5050**, pp 262-292, (2008)

[23] Yaochu Jin, J.Branke. *Evolutionary Optimization in Uncertain Environments– A Survey*. IEEE Transactions on Evolutionary Computation **9**(3), 303-317 (2005)

[24] Michel Gendreau, Gilbert Laporte, René Séguin. *Stochastic vehicle routing problem*, European Journal of Operational Research **88**(1), pp 3-12 (1996)

[25] Eshetie Berhan, Birhanu Beshah, Daniel Kitaw. *Stochastic Vehicle Routing Problem: A Literature Survey*, Journal of Information Knowledge Management **13**(3), 12 (2014)

An inventory location routing model with environmental considerations

Sanaa Lerhlaly[1,2,*], *Maria* Lebbar[3], *Hamid* Allaoui[1], *Sohaib* Afifi[1], and *Driss* Ouazar[2,3]

[1] LGI2A, Université d'Artois, France
[2] LASH, Ecole Mohammadia d'ingénieurs Université Mohammed V de Rabat .Morocco
[3]SYSCOM, Ecole Nationale Supérieure des Mines de Rabat, Morocco

Abstract. In this work we consider an integrated Inventory Location Routing problem ILRP. The objective is to select depots to open, the routes to perform distribution from the capacitated depots to the clients by a homogeneous fleet of vehicles and over a multi-period planning horizon. To solve the problem, an exact model is designed in order to minimize the total cost (location, routing and inventory) and the routing CO2 emissions, while respecting transport and storage constraints for a hazardous material. We have also studied an industrial real case to illustrate the problem.

1 Introduction and literature review

The design of efficient supply chains has become more challenging in view of environmental regulations, especially for Hazardous materials (Hazmat) supply chains. The decisions for supply chain management can be categorized into three types: strategic, tactical, and operational. The classical approach for the optimization of logistics systems is the decomposition of the system into sub problems according to these decision levels. Facility location problems (strategic level) deal with defining the number, the location and the capacity of the warehouses, in order to meet customer demand at lower cost. This problem was reviewed by [1-3]. The vehicle routing problem (VRP) (tactical or operational level) was studied in [4-7]. The solving techniques used for this problem were summarized by [8-10].

This approach has many advantages such as simplification of modeling and problem solving techniques. This can, though, lead to a sub optimality of solutions, because it doesn't take account of the entire problem [11].

In order to a global view of the supply chain design, some papers integrate the three components: location, routing and inventory. The problem becomes more complex, but it is important to anticipate interactions between decisions levels.

The Inventory Location-Routing Problem (ILRP) can be considered as an approach for integrating the three components (depot location, vehicle routing and inventory management) into one model, in order to optimize the supply chain design and minimize the total costs. In [12,13], authors presented a location-routing problems (LRP) including nonlinear inventory costs in the objective function. They consider a single product and multi-depot with unlimited capacity and stochastic demand. In order To solve this problem, Lui and Lee [12] used a two-phase heuristic method, while Liu and Lin [13] proposed a hybrid heuristic based on the Tabu Search (TS) and the Simulated Annealing (SA). Another nonlinear stochastic model was studied by Shen and Qi [14] in order to integrate the location, inventory, and the routing approximate costs. Javid and Azad [15] presented a stochastic model in order to find simultaneously the decisions of location, allocation, inventory at depots and routing. The authors proposed an exact solution and then a heuristic method based on hybridization of Tabu Search (TS) and Simulated Annealing (SA). Regarding the deterministic demand, Ambrosino and Scutella [16] used a linear programming relaxation for the ILRP. Hiassat and Diabat [17] studied an inventory location routing problem with deterministic demand for perishable product. They proposed a multi-period model formulated as Mixed Integer Program (MIP). Recently, Guerrero et al. [18] presented an algorithm for a multi-period ILRP with deterministic demand. They proposed a hybrid approach in order to optimize simultaneously location of depots, allocation of retailers to depots, inventory at both depots and retailers, and multi-period routing. The algorithm optimizes globally the components of the problem by sharing information between decisions levels. Guerrero et al. [19] investigated a Relax and Price heuristic for the same problem. They coordinated the generation of new columns in the pricing problem and updated Lagrangian multipliers. Results show important cost savings and efficient computation if compared to the previous heuristics. Ying Zhang et al. [20] proposed a hybrid metaheuristic solution in order to solve an ILRP. They considered a multiple depots and geographically dispersed customers. Nekooghadirli et al. [21] studied a bi-objective location-routing-inventory (LRI) model

* Corresponding author: sanaalerhlaly@gmail.com

considering multi-period, stochastic demands, multi-product and probabilistic travelling time among customers. Authors presented a mixed integer nonlinear programming model in order to minimize the total cost and maximize mean time for delivering commodities to customers.

This paper deals with an integrated Inventory Location Routing Problem (ILRP) considering a deterministic demand and a multi-period horizon of decisions. In order To take into account risks related to the use of Hazmat, we consider storage constraints at depots and clients, small capacity vehicles, and routes with fixed number of customers which must not be exceeded.

2 Problem definition

We are studying a supply chain design problem for Hazmat. We have to distribute a hazardous product from a single manufacturer to a set of customers, through a set of intermediate depots with homogeneous fleet. This aims to minimize both of the total cost and the routing CO2 emissions. The total costs signify the depots opening, the routing, and the clients inventory costs.

A first model for this problem was presented in S.Lerhlaly et al. [22].The problem is formulated as a set covering model whose the decision variables are the routes. A case study is also provided.

Let: (*I*) denote the set of (*m*) candidate depots. (*J*) denote the set of (*n*) customers, (*T*) denote the set of (*p*) periods in the planning horizon, and (*K*) denote the set of vehicles. $V = \{J \cup I\}$ is the set of nodes in the graph $G = (V;\ A;\ C)$. C is the cost matrix (c_{ij}) associated to the traveling cost from node i to node j in the set of arcs A.

The hazardous materials regulations limit the storage and the transportation quantities which increase the transportation frequency, and augment the environmental impacts due to CO2 emissions. That is why it is important to add these environmental impacts to the objective function. The estimation of CO2 emission for transportation requires complex calculations, because of the difficulty of quantifying some variables such as driving style, weather conditions, congestion...[23,24]. In order To calculate the emissions factor CO2, We are using a method similar to that used by S. Ubeda et al. [25].

2.1 Parameters

O_i Opened cost of depot $i \in I$
f_{jt} Inventory holding cost of customer $j \in J$ in time period $t \in T$
F vehicles cost
Cap Capacity of vehicule K
d_{jt} Demand of customer $j \in J$ in time period $t \in T$
S_{j0} Inventory level at customer $j \in J$ at the beginning of time period $t = 1$
b_j Maximum inventory level at customer j
Cpd_i Maximum storage capacity at depot i
MaxC Maximum number of customer by route
E_j CO2 emissions factor
$\alpha,\ \beta,\ \lambda,\ \Omega$ Weights factor associated with opening cost, transportation cost,CO2 emission ,inventory cost.

2.2 Decisions variables

$y_i = 1$ if depot $i \in I$ is opened and 0 otherwise
$x_{ijkt} = 1$ if the arc (i,j) used in the route performed by the vehicle $k \in K$ on period $t \in T$ and 0 otherwise
$f_{ij} = 1$ if clients j is assigned to depot I and 0 otherwise
S_{jt}= Inventory holding of customer $j \in J$ in the end time period $t \in T$
Q_{ijkt}= Quantity delivered from depot $i \in I$ to customer $j \in J$ by vehicle $k \in K$ in time period $t \in T$.

2.3 Model

$$\min \alpha \sum_{i \in I} O_i y_i + \beta \sum_{i \in V} \sum_{j \in V} \sum_{k \in K} \sum_{t \in T} c_{ij} x_{ijkt} + \sum_{i \in I} \sum_{j \in J} \sum_{k \in K} \sum_{t \in T} F\, x_{ijkt} + \lambda \sum_{i \in V} \sum_{j \in V} \sum_{k \in K} \sum_{t \in T} c_{ij} e_j x_{ijkt} + \Omega \sum_{j \in J} \sum_{t \in T} m_{jt} S_{jt} \quad (1)$$

Subject to

$$\sum_{i \in I} \sum_{j \in J} Q_{ijkt} \le Cap \qquad \forall k \in K, \forall t \in T \quad (2)$$

$$\sum_{j \in V} x_{ijkt} - \sum_{j \in V} x_{jikt} = 0 \qquad \forall k \in K,\ \forall t \in T,\ \forall i \in V \quad (3)$$

$$\sum_{i \in j} \sum_{k \in K} x_{ijkt} \le 1 \qquad \forall j \in J,\ \forall t \in T \quad (4)$$

$$\sum_{i \in j} \sum_{k \in K} x_{jikt} \le 1 \qquad \forall j \in J,\ \forall t \in T \quad (5)$$

$$\sum_{i \in I} \sum_{j \in J} x_{ijkt} \le 1 \qquad \forall k \in K,\ \forall t \in T \quad (6)$$

$$\sum_{i \in I} f_{ij} = 1 \qquad \forall j \in J \quad (7)$$

$$\sum_{a \in J} x_{iakt} + \sum_{a \in V} x_{ajkt} \le 1 + f_{ij} \qquad \forall j \in J, \forall i \in I, \forall k \in K, \forall t \in T \quad (8)$$

$$\sum_{j \in J} \sum_{k \in K} Q_{ijkt} \le cpd_i * y_i \qquad \forall i \in I,\ \forall t \in T \quad (9)$$

$$S_{jt-1} + \sum_{i \in I} \sum_{k \in K} Q_{ijkt} = d_{jt} + S_{jt} \qquad \forall j \in J,\ \forall t \in T \quad (10)$$

$$S_{jt} \le b_j \qquad \forall j \in J,\ \forall t \in T \quad (11)$$

$$\sum_{i\in V}\sum_{j\in V} x_{ijkt} - 1 \le MaxC \qquad \forall t\in T,\ \forall k\in K \quad (12)$$

$$\sum_{j\in S}\sum_{j\in S} x_{ijkt} \le |S| - 1 \qquad \forall S\in J, \forall k\in K, \forall t \quad (13)$$

$$\sum_{t\in T}\sum_{k\in K} Q_{ijkt} \le M * f_{ij} \qquad \forall i\in I,\ \forall j\in J \quad (14)$$

$$Q_{ijkt} \le M * \sum_{u\in J} x_{iukt} \qquad \forall i\in I, \forall j\in J, \forall t\in T, \forall k\in K \quad (15)$$

Objective function (Equation1) includes the sum of location, transportation, inventory costs and CO2 emissions. The first term represents the opening depots fixed cost, the transportation costs are minimized by second and third terms, CO2 emissions costs is minimized by the fifth term and the last one represents the inventory holding cost. Constraint (2) indicates the capacity constraints of vehicles. Constraints (3-5) present vehicle flow conservation. Constraint (6) forces each vehicle to perform one route per period at the most. Constraint (7) guarantees that each customer must be allocated to a single opened depot. Constraint (8) state that a retailer j can't be linked to a depot i if j is not assigned to depot i. The capacity of each depot is presented in Equation (9).Equation (10) represents the inventory balance. Equation (11) ensures that the inventory level at a customer never exceeds maximum storage limits due to the Hazmat regulations. Equation (12) respects the maximum number of customers per route. The traditional subtour elimination constraints are presented in Equation (13). Constraints (14) (15) guarantee that if a retailer is served on period t with route k, it must be visited.

To validate the proposed model, a real case study has been developed in order to optimize the design of chlorine supply chain (for a leading company in the drinking water distribution sector in Morocco). This company use chlorine for water disinfection. The problem considers 10 delivery points (Clients).The maximum number of clients per route is 3.The Number of candidate depots is 5. The planning horizon is three time periods.

Table 1. Simulation results

Eco2 weight	Total cost	D.C	N.D	R.C	I.C	Eco2
λ=0	718498,6	700000	1	7458	1040,6	3898,7
λ=100	719287,1	700000	1	7334	1953,1	3586,4
λ=1000	2156245,6	2140000	3	4537	1708,6	2862,2
λ=10000	2936148,1	2920000	4	4438	1710,1	2026,7

The results of simulations are reported in Table1. The first column of this table indicates the weight of the CO2 emissions. The second column presents the total cost. Columns: D.C, N.D, R.C, I.C respectively indicate the costs and number of opened depots, the routing and inventory costs. The last column gives the CO2 emissions quantity.

The results show that the total cost and the number of selected depots increased as the weight of the transportation CO2 emissions increased (Fig.1, Fig.2).

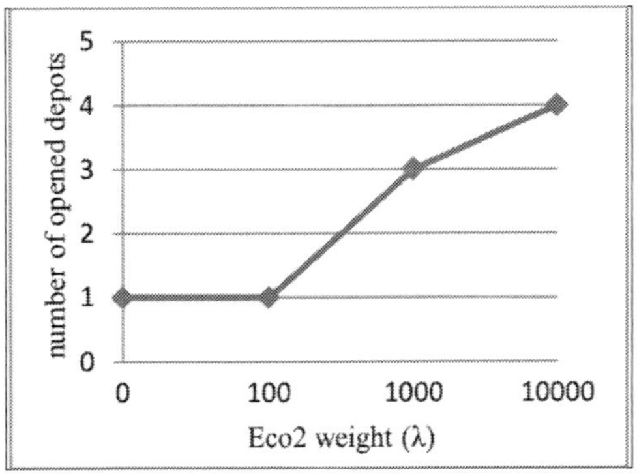

Fig. 1. Number of opened depots vs Co2 emissions weight

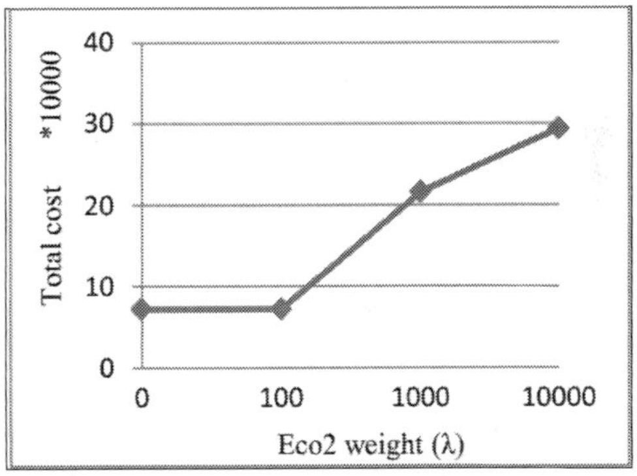

Fig. 2. Total cost vs Co2 emissions weight.

3 Conclusion

We present an inventory location routing problem (ILRP) that optimizes the operating costs, and routing's CO2 emissions taking into account Hazmat constraints. We modeled the problem as an exact integer linear program. The developed model is applied to a case study of Hazmat supply chain to demonstrate its applicability. Preliminary results are promising for medium size instances. The future work is both the development of a heuristic method, based on a guided generation of routes, and also the testing of some larger scale instances.

References

1. Z. Drezner, Springer, New York, (1995).
2. S.H. Owen, M. S. Daskin, European Journal of Operational Research, (1998), **111(3),** 423-447.
3. M.T. Melo, S.Nickel, F. Saldanha-da-Gama, European Journal of Operational Research, (2009), **196(2),** 401-412.

4. Bodin, L.D., Golden, B.L., Assad, A.A., and Ball, M.O. Computers and Operations Research (1983), **10**, 69-211.
5. G. Laporte, H.Osman, Annals of Operations Research (1995), **61**, 227-262.
6. G. Laporte. Transportation Science, (2009), **43(4),** 408-416.
7. M. Drexl, European Journal of Operational Research, (2013), **227(2),** 275 -284.
8. G. Laporte, (European Journal of Operational Research 1992), **59**, 345-358.
9. G. Laporte, M. Gendreau, J. Potvin, F.Semet, International transactions in operational research, (2000), **7(4-5),** 285-300.
10. P. Toth, D. Vigo, Discrete Applied Mathematics, (2002), **123(1-3)**, 487-512.
11. C. Prodhon, *PhD Thesis*, University of Technology of Troyes, France, (2006)
12. S.C. Liu, S.B. Lee, The International Journal of Advanced Manufacturing Technology, **22** (2003), 941–950.
13. S. Liu, C. Lin, The International Journal of Advanced Manufacturing Technology, **26** (2005), 372-381.
14. Z.J.M. Shen, L. Qi, European Journal of Operational Research, **179** (2007), no. 2, 372–389.
15. A. Ahmadi Javid, N. Azad, Transportation Research Part E: Logistics and Transportation Review, **46** (2010), no. 5, 582 - 597.
16. D. Ambrosino, M.G. Scutell`a, European Journal of Operational Research **165** (2005), no. 3, 610–624.
17. A. Hiassat, A. Diabat, *Proceedings of the 41st International Conference on Computers and Industrial Engineering*, (2011).
18. W.J. Guerrero, C. Prodhon, N. Velasco, C.A. Amaya, International Journal of Production Economics, **146** (2013), no. 1, 359–370.
19. W.J. Guerrero, N. Velasco, C. Prodhon, C.A. International Transactions in Operations Research, 22 (2015), 129–148.
20. Zhang et al., 2014, Y. Zhang, M. Qi, L. Miao, E. Liu, Transport. Res. Part E: Logist. Transport. Rev., **70** (2014), pp. 305–323
21. Nekooghadirli, N., Tavakkoli-Moghaddam, R., Ghezavati, V., Javanmard, S. Comput. Indust. Eng (2014)., **76**, 204–221.
22. Sanaa Lerhlaly, Maria Lebbar, Hamid Allaoui, Driss Ouazar, Sohaib afifi, Contemporary Engineering Sciences, Vol. **9**,(2016) ,no.7,303-314.
23. A. Palmer, *Logistics Research Network Conference Proceedings,* University of Hull, Hull, (2007), 27–32.
24. T.V. Van Woensel, R. Creten, N. Vandaele, Production and Operations Management, **10** (2001), 207–223.
25. S. Ubeda, F.J. Arcelus, and J. Faulin, International Journal of Production Economics, **131** (2011), 44-51.

MATEC Web of Conferences **105**, 00003 (2017) DOI: 10.1051/matecconf/201710500003
IWTSCE'16

Communal analysis of the tramway accidentology in France

Adnan Bentama[1,*], Marine Millot[2] and Abdellah Khatory[1]

[1]*Laboratory of Production engineering, Energy and Sustainable Development, High School of Technology, Sidi Mohamed Ben Abdellah University, Fez, Route d'Imouzzer BP : 2427, 30 000, Morocco.*
[2]*CEREMA Méditerranée, DAT-ARTU, Pôle d'activités des Milles, 13593 Aix-en-Provence, cedex 3, France.*

Abstract. Introduction: The purpose of this study is to assess the impact of communes size on the number and severity of tram accidents in France. **Methodology:** For this we calculated the average densities of accidents and serious victims of trams per line kilometer for each class of commune between 2012 and 2014. While adverting to the official data of accidentology and population. **Results:** The study has shown that almost one third of the municipalities served by the tram have known no accident. The study also showed that the size of the communes is proportional to the accidents density per kilometer of the tram line. The study has also concluded that the density of serious victims per line kilometer is maximum in the communes of average size. **Conclusion:** To carry out the modal shift toward the urban public transport, it is necessary to deepen our knowledge of the specificities of insecurity of these modes. This study is part of this framework, and made the link between the size of the communes and trams accidentology in France.

1 Introduction

The tramway is among the transport systems most used in French agglomerations. Indeed, France has 28 tram networks, and as they are among means of public transport with a character, mainly, urban, they are always in interaction with the other road users, this sharing causes, sometimes, accidents which are, in certain cases, serious and very mediatized.

However, few researches were interested in the impacts of public transport in term of road safety [1]. Certain authors explain it by the low percentage of accidents involving public transport, but likewise by the idea that public transport having to contribute to a reduction of the role of the automobile, and this must have a positive event on safety [2]. Others further go, by considering that the public interest in road safety of this mode is not as large as that of the other types of vehicles, the level of safety of public transport being judged like acceptable [3].

Nevertheless, accidents involving trams (all networks), and the resulting casualties are far from being negligible. Indeed, the three years of the study (2012 to 2014) have registered: 438 accidents involving trams which have caused 538 victims.

It is therefore important to take these problems of road safety into account, particularly in the recent context of reintroduction of trams in many French cities. After a period of gradual dismantling of tram networks. But in the past few decades the tramway has made its "comeback" in the heart of French urban centers, playing an important role in urban policies [4, 5]. In summation to this role, today, it is a central tool of transportation policy, to answer to the rising need for mobility and to guide the governance and the urban functioning by promoting downtown and strengthening territorial cohesion.

The works which were concerned with the aspect of tram safety, established that the tram pose specific safety problems, by direct collisions, for the vulnerable users in particular pedestrians [6, 7, 8]. The tram can intervene, also, indirectly in the genesis process of accidents, for example: A user who crosses the road masked by a tram and hit by a car [9, 10, 11, 12]. The publications which examined specificities of tram accidents, in particular the types of involved thirds, show that the tram passengers, the pedestrians and the car passengers are generally the main victims of trams [13, 14]. Other researchers have interested on the impact of the configuration of public spaces redesigned for trams on road users safety. They conclude that this configurations may have an impact on the occurence of accidents, even when the tram vehicle is not present.[15, 16, 17]

The question that we posed in this paper is linked, primarily, to the congestion at habitat environments of dense and continues nature, which increases the probability of collisions between tramways and the other road users. This approach of the density shows the

* Corresponding author: adnan.bentama@usmba.ac.ma

 DOI: 10.1051/matecconf/201710500003

complexity of the reflection. In effect, it carries on the effect of an urbanization more dense, which could reduce the severity of accidents, but conversely could increase their frequency [18]. We are going to try, through this work, to check this assumption for the tram system, and find out what is the impact of the communes size on the accidents number and the victims severity of the tram system in France. To do this, have been calculated the mean densities of accidents and serious victims of trams per line kilometer for each class of commune in France between 2012 and 2014. While adverting to the official data of accidents and population.

2 Data and methodology

2.1 Data

The work is conducted from the Analysis Balance of Road Traffic Injury Accidents (BAAC), which is a French national basis established by the order forces, from the minutes made as soon as an accident occurs and involving at least one victim. This database includes detailed information on the characteristics and the general circumstances of the accident occurrence (time, places, vehicles and persons involved, etc.). The BAAC files are coupled to GIS software (Geographic Information System) named CONCERTO, this software allows the statistical and cartographic treatment of data.

The duration of the study is spread over three years from 2012 to 2014, this period has recorded, according to the BAAC files, 175440 accidents (all modes) distributed throughout the territory of the Metropolitan France. Our study will treat just tram accidents, and this on all the territory of the Metropolitan France.

2.2 Methodology

To deeply analyze the accidents of the tram per communes, and to avoid the comparisons between raw data, we tried to calculate the average density of accidents and victims per km of tram line for each class of commune. Indeed, we managed to have the layout of tram lines in all the networks of tram in the territory of Metropolitan France, after that we use a communal map of France, Next we could calculate the length of tram networks line by each commune by using MAP-info software, we gathered these communes by size, relating to the population of each commune by using the official statistics of the National Institute of Statistics and Economic Studies (INSEE) of 2013. For accidentology data, CONCERTO enabled us to have the number of accidents and victims of tram in each commune. After having obtained for each commune the tram accidents data, the layout length of tram lines and the population data, we could calculate the average density of tram accidents and the serious victims for each class of commune.

The diagram below recapitulates the methodology used in this work:

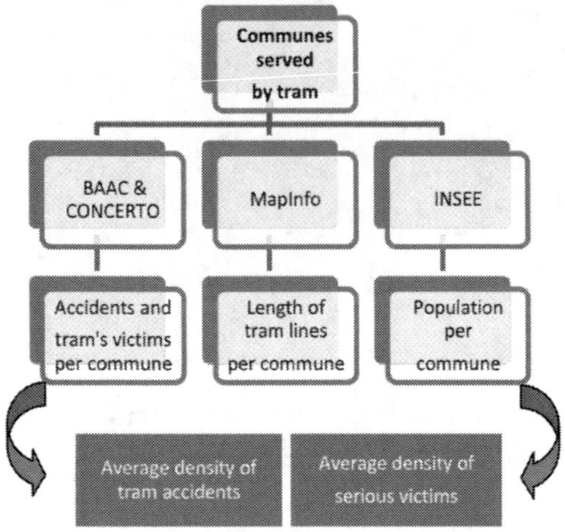

Figure 1. Descriptive diagram of the methodology followed for the analysis of the tram accidents per communes

3 Results

3.1. Communes without accidents

Table 1 Communes without tram accidents per size in France from 2012 to 2014

Communes size (Thousands inhabitants)	2–5	5–20	20–50	50–100	100–300	>300
Number of communes served	13	52	61	14	20	5
Number of communes served without accidents	6	21	19	4	1	1
Percentage of communes without accidents	46,15	40,38	31,15	28,57	5,00	20,00

On the 165 communes served by the trams in France, 52 communes did not know any accident of tram during the three years of the study (see table 1), they concern in the majority of the communes whose populations lie between 5 and 50 thousand inhabitants.

Note: It is noted that we did not take into account the tram networks which were operational after or during the study period, therefore we took just the operational networks before 2012, i.e. we didn't consider the tram networks of Le Havre, Tours, Besancon, Dijon and Brest. The tram networks included in the study are: Paris, Toulouse, Nice, Lyon, Nantes, Marseilles, Strasbourg, Montpellier, Bordeaux, Lille, Reims, Saint-Etienne, Grenoble, Angers, Mans, Clermont-Ferrand, Orleans, Rouen, Mulhouse, Caen, Nancy, Valenciennes and Douai.

3.2 Average densities of accidents and serious victims

Table 2 Annual average density of the accidents and the serious victims of the tram per kilometer of line according to communes size in France from 2012 to 2014

Communes size (Thousands inhabitants)	Length of lines (km)	Number of accidents	Number of serious victims (Killed + HI)	Annual average density of the accidents/Km of line [a]	Annual average density of serious victims /Km of line [a]
2--5	11,23	*1*	*0*	*0,03*	*0*
5--20	63,28	25	*6*	0,13	*0,03*
20--50	131,2	70	36	0,18	0,09
50--100	25,28	20	*10*	0,26	*0,13*
100--300	368,7	219	74	0,2	0,07
>300	66,6	67	*15*	0,34	*0,08*

[a] densities where the number of accidents or serious victims are under 20 are noted in italic.

The examination of tram accidents in France per class of commune show that communes from 100 to 300 thousand inhabitants record the most tram accidents with 219 accidents, in second position we find communes from 20 to 50 thousands inhabitants and the communes of 300 thousands inhabitants with a very close numbers of accidents (70 and 67 accidents respectively), in third position come communes from 5 to 20 MH and the communes from 50 to 100 MH also with practically the same number of accidents (25 and 20 accidents respectively).

➢ **Average density of tram accidents per class of commune**

The communal analysis of the average density of tram accidents in France, shows that the communes of more than 300 thousand inhabitants which have the ratio more raised. Indeed, this type of commune recorded 0,34 accident for each kilometer of tram line annually. Followed by the communes of 50 to 100 thousand inhabitants, with a ratio of 0,26 accident per kilometer of line per annum. In third place we find communes of 100 to 300 thousand inhabitants with a ratio of 0,2 accident per km of tram cable. The communes of 20 to 50 thousand inhabitants knew an annual density of tram accidents of 0,18 accident per kilometer of tram line. Lastly comes the communes from 5 to 20 thousand inhabitants with a ratio of 0,13 accident/Km of the tram line.

➢ **Average density of serious victims of tram per class of commune**

The analysis of the annual average density of serious victims per class of commune between 2012 and 2014, reveals that the communes of 50 to 100 thousand inhabitants record the highest average of serious victims density, with a ratio of 0,13 serious victim per km of tram line. Followed by communes of 20 to 50 thousand inhabitants, the communes of more than 300 thousand inhabitants and the communes of 100 to 300 thousand inhabitants with very close ratios, indeed these three commune classes recorded (respectively) 0,09, 0,08 and 0,07 serious victim per km of line per annum. Lastly come the communes from class 5 to 20 thousand inhabitants with a ratio of 0,03 serious victims per km of tram line annually. However, it should be noted that it is necessary to interpret these results with precaution, because of the weak frequencies of serious victims in some classes of communes.

4 Discussion

This study, which is spread out over three years from 2012 to 2014, aimed to analyze the impact of communes size on tram accidents in France. By evaluating for each class of commune two parameters which are; the average density of tram accidents and the average density of the serious victims per km of tram line.

The study showed that the communes of 100 to 300 thousand inhabitants which knew the majority of tram accidents in France. But this result cannot enable us to conclude that these communes are overrepresented, because it is necessary to relativise these figures and to cross them with an indicator as the number of kilometers of tram line.

The analysis showed that the communes of big size (+300 thousand inhabitants) which record the most raised annual density of accident per km of tram line. An expected result, which confirms the starting assumption which says that; the densest habitat environments record the most accidents. Because of the strong frequentation of the road by all types of users, what specifically increases the probabilities of collision with the trams. Broadly speaking, it is observed that the annual average of density of the tram accidents per kilometer of line in France is proportional to the size of the communes except communes of 100 to 300 thousand inhabitants. Indeed, this type of commune recorded a density of accident of tram less than the communes of 50 to 100 thousand inhabitants.

Concerning the severity, the study showed that the average density of serious victims caused by tram accidents tram is higher in the communes of 50 to 100 thousand inhabitants, that can be explained by significant speeds practiced by tram drivers in the habitat environments of less density. In order to increase the commercial speed of the transport system, which causes serious accidents, because the tram as being a heavy vehicle on rails it requires a long braking distance, in particular if the speed is significant. But it is necessary to take these results with precaution because of the weak numbers of serious victims in some classes of communes. The explanation of speed can be reinforced by the fact that the two types of communes which recorded the most serious victims (36 serious victims for communes with 20 to 50 thousand inhabitants and 74 serious victims for communes with 100 to 300 thousand inhabitants) where it is noticed that the communes of size 20 to 50 thousand inhabitants recorded an annual density of serious victims per km of line of tram higher than that of the communes of 100 to 300 thousand inhabitants.

Some limit must be evoked concerning this study. Firstly, the duration of the study, if the duration was 5 years, that would be better. Especially when we work on a displacement mode, like the tram, which does not make too many accidents. Secondly, weak figures of the data in particular serious victims which poses some problems of statistical significance.

5 Conclusion

To conduct the modal transfer towards Urban Public Transport and to decrease the use of the private car, first of all, it is necessary to improve safety of these modes via a deep comprehension of its insecurity specificities. This study aimed to evaluate the impact of the size of communes on tram accidents in France between 2012 and 2014. The study concluded that the size of the communes is proportional to the accidents density per kilometer of tram line. The study also showed that the density of the serious victims per kilometer of line is maximal in the communes of average size.

References

1. Brenac T., Clabaux N. (2005) The indirect involvement of buses in traffic accident processes. Safety Sci 43:835-843
2. Beer S., Brenac T. (2006). Tramways et sécurité routière, l'expérience des pays germanophones. Transp Environ Circ 190:40-46
3. Cafiso S., Di Graziano A., Pappalardo G. (2013) Road safety issues for bus transport management. Accid Anal Prev 60:324-333
4. Frenay, P. (2005). Le tram, outil en faveur d'une ville durable?: Réflexions tirées d'une comparaison entre quelques villes moyennes françaises et alémaniques ». TEC, 185, 2-8.
5. González, Rubén C. Lois, Miguel Pazos Otón, and Jean-Pierre Wolff. "Le tramway entre politique de transport et outil de réhabilitation urbanistique dans quelques pays européens: Allemagne, Espagne, France et Suisse."Annales de géographie. No. 6. Armand Colin, 2013.
6. Hedelin, A., Björnstig, U., & Brismar, B. (1996). Trams—a risk factor for pedestrians. Accident Analysis & Prevention, 28(6), 733-738.
7. Hedelin, A., Bunketorp, O., & Björnstig, U. (2002). Public transport in metropolitan areas—a danger for unprotected road users. Safety science,40(5), 467-477.
8. Isabelle Chalanton et Muriel Jadoul. (2009). « Etude des accidents entre un Tram et un piéton en région de Bruxelles capitale », Institut Belge pour la Sécurité Routière (IBSR).
9. Schnabel, W., Korn, J. (2001). Accident occurence in tram traffic. Proceedings of the international conference 'Road safety on Three Continents' (Pretoria), Staten's VAEG-OCH Transport for sknings instiut, n. 15(A), pp. 516-526.
10. Unger, R., Eder, C., Mayr, J-M., Wernig, J. (2002). Child pedestrian injuries at tram and bus stops. Injury, n.33(6), pp. 485-488.20.
11. Kobi, R. (1990). Sicherheit an Bus-und Tramhaltestellen. Zeitschrift fur Verkehrssicherheit, n. 36(2)
12. Krusyzyna, M., Rychlewski; J. (2013). Influence of approaching tram on behaviour of pedestrians in signalised crosswalks in Poland. Accident Analysis & Prevention, 55, pp. 185-191.
13. BENTAMA, A., & KHATORY, A. (2015). Etude statistique des risques du tramway Rabat-Salé/Maroc.11eme Congrès International de Génie Industriel, Québec les 26-28 octobre 2015.
14. Amon, B. (2002). Unfälle mit Straßenbahnen. In : Hauger, G. (Ed.), Forum Verkehrssystemplanug 2002, Ausgewählte Seminararbeiten, IVS-Schriften Band 15, Institut für Verkehrssystemplanung der TU Wien, Österreichischer Kunst- und Kulturverlag, Wien. Pp. 7-24.
15. Cameron, I. C., Harris, N. J., & Kehoe, N. J. S. (2001). Tram-related injuries in Sheffield. Injury, 32(4), 275-277.
16. Kobi, R. (1990). Sicherheit an Bus-und Tramhaltestellen. Zeitschrift für Verkehrssicherheit, 36(2).
17. Maitre, E. (2015). Public spaces re-designed for trams in French cities: safety concerns. Advances in Transportation Studies, (37).
18. Millot, M. Etude des liens complexes entre formes urbaines et insécurité routière. Rapport d'étude, CERTU, 2004.

MATEC Web of Conferences 105, 00004 (2017) DOI: 10.1051/matecconf/201710500004
IWTSCE'16

Conformity checking of LPG Transportation Trucks by Modelling and Simulation

Maryam Gallab[1,3,*], *Hafida* Bouloiz [2], *Mohamed* Tkiouat[1] and *Emmanuel* Garbolino[3]

[1]LERMA, Engineers Mohammadia School, Mohammed V University, Avenue Ibn Sina BP765, Agdal, Rabat, Morocco.
[2]Industrial Engineering Department, LG2I, National School of Applied Sciences, Ibn Zohr University, - B.P: 1136 Agadir – Morocco.
[3] Crisis and Risk research Centre (CRC), Mines Paris-Tech, 1 rue Claude Daunesse CS 10207 06904 Sophia Antipolis Cedex France.

Abstract. LPG Loading /unloading site is considered as a dangerous environment of significant risk, especially during LPG transfer operations. These risks may cause catastrophic dangers such as BLEVE (Boiling Liquid Expanding Vapour Explosion), UVCE (Unconfined Vapour Cloud Explosion), etc. The site consists of a tank (RST) containing LPG and three transfer posts. Two posts allow loading of small carrier tankers 6 tons. The other one transfer post transfer LPG contained in jumbo tankers 20 tons inside the tank. The industrial site at risks chooses to demonstrate the need for a modelling-simulation approach. The aim is to check first compliance and authorization of the truck when it enters the site for loading / unloading, to verify the authorization to make the transfer, and then, to identify the driver and his training to ensure that the transfer operations are going to take place without incident. Sometimes, it may happen that Consignment Operator (CO) is busy or absent and thus the driver operates alone without being authorized, in this case the site risks a loss of control due to a lack of resources. The modelling-simulation will be done using multi-agent systems that will present the drivers, trucks, PO and loading / unloading posts as agents in order to have a model facilitating this checking.

1 Introduction

LPG (Liquefied Petroleum Gas) is considered as a by-product of the petrochemical industry, and has become since the 1930s, a powerful energy source [1].
LPG domain has a long history of technological advances and major accidents.

Indeed, LPG sector is considered among the most hazardous industrial sectors based on accidents occurrence statistics. Among these accidents:

- Feyzin in France (1966),
- San Juan Ixuatepec disaster in Mexico (1984),
- Izmit in Turkey (2002).

Latest accidents that touched LPG transportation are centralized on the ARIA database (Research and Information Analysis accidents) such as Bollene accident (2008), Morannes accident (2011), Bastia accident (2014), and Lamentin (2015).

The aim is to check first compliance, conformity and authorization of the truck when it enters the site for loading / unloading, to verify the authorization to make the transfer, and then, to identify the driver and his training to ensure that the transfer operations are going to take place without incident in order to avoid leaks and problems leading to explosions.

This paper is structured as follows: Section 2 presents LPG Loading/Unloading Site. Section 3 shows the hazards and accidents that may generate from non-verification of the trucks and drivers. In section 4, the proposed model and simulation results are given. Finally, a conclusion and perspectives are provided.

2 LPG Loading/ Unloading Site

The field of study is the LPG loading/unloading site. LPG sector in France represents approximately 6,000 industrial sites. These sites are divided into: Deposits bottles, filling centres for filling bottles, and bulk relays which are loaded tankers supplying the customer tanks.

The site assures its supply by jumbo tankers from suppliers (Lavera) and distributes LPG to these customers by small carrier tankers. The site consists of a propane mounded tank, transfer equipment in the pump station area (pump / compressor / piping), a loading / unloading post of jumbo tankers and two loading post of small carrier tankers, and a parking area for tankers and bottles [2, 3].

* Corresponding author: meryam09@gmail.com

DOI: 10.1051/matecconf/201710500004

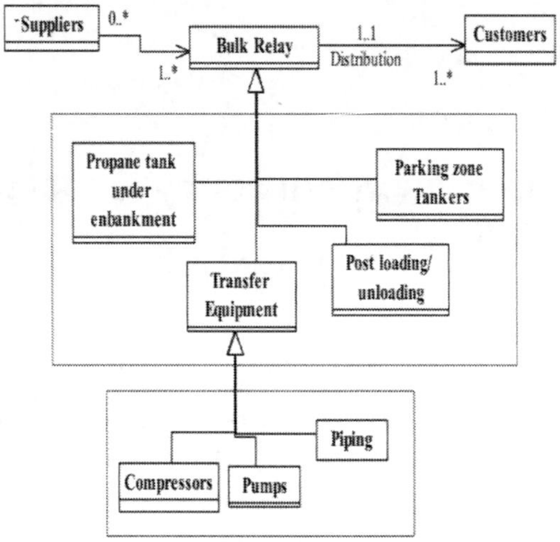

Fig. 1. LPG Loading/Unloading Site.

2.1. Truck Control Points

Before the trucks enter the site and perform the loading / unloading operations, a control is carried out at the site entrance by a Consignment Operator (CO) to check the following points:

- Check the driver identification,
- Check permission to load/unload,
- Check the conformity of the tank,
- Check compliance with the quotas,

If all these points are checked and compliant, a CO provides a bill of loading to the driver, this driver heads towards the transfer station.

2.2 Checking the driver identification

The site is in "self-service", drivers perform manipulations to load / unload their trucks. Although everything is automated (depending on the tank, the system knows exactly how many LPG must be injected or aspirated).

A Consignment Operator (CO) monitors operations, especially when unloading jumbo tankers.
The system counts about 400 drivers who can be:

- Trained and audited ➔ they can operate in self-service,
- A driver is simply formed ➔ he operates under the CO supervision,
- The driver is not formed ➔ the CO who will operate and the driver observes.

3 Hazards generated

The problem is that failure to verify these points can lead to explosions such as:

- **BLEVE** (Boiling Liquid Expanding Vapor Explosion): Explosive vaporization of a boiling liquid is a scenario analogous to an explosion induced by the rapid expansion of flammable vapors produced by a gaseous substance preserved under pressure in liquid phase in a medium confined; Of this event, can occur effects of overpressure and thermal irradiations (fireball) for people and structures.

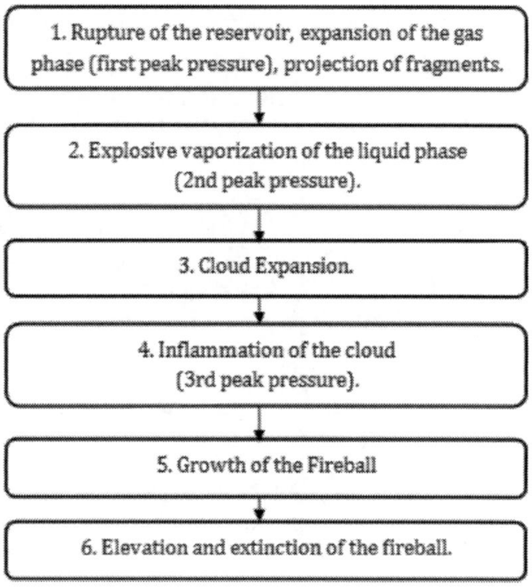

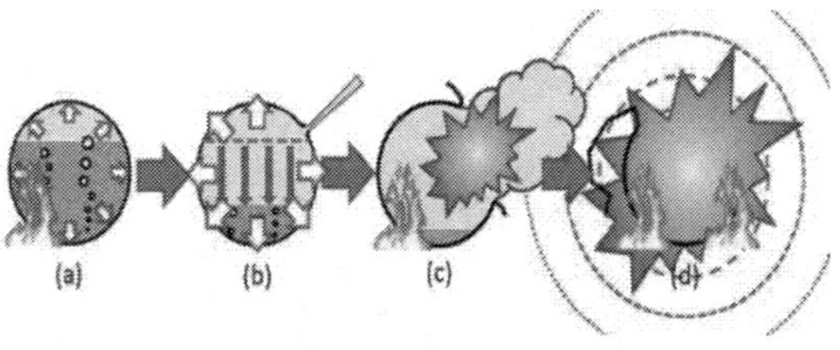

Fig. 2. BLEVE Formation.

- A **UVCE** (Unconfined Vapour Cloud Explosion) is an accident scenario induced by the reject and the dispersion of flammable substance gas or vapour phase in an unconfined environment, from which may arise, in case of source ignition, thermal and overpressure effects for Humans, structures and the environment. From this explosion, will produce thermal and overpressure effects which depend on local conditions and, specifically, weather and gas mixture conditions.

A UVCE, generally, comprises the following steps:

DOI: 10.1051/matecconf/201710500004

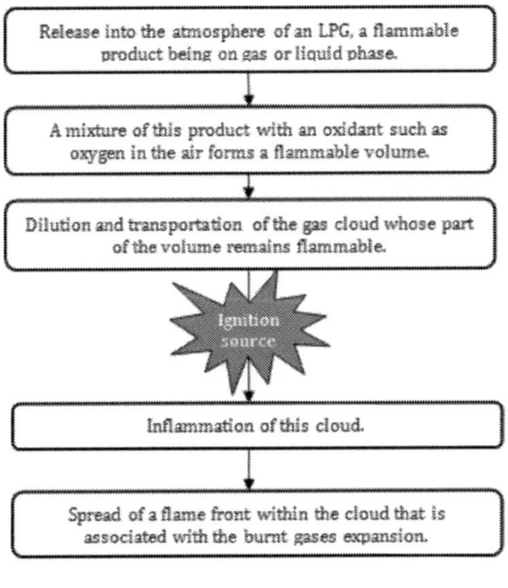

Fig. 3. UVCE Formation.

4 The proposed Model

To avoid dangers and to ensure a healthy environment for loading / unloading operations, we propose a model to facilitate verification based on Modelling and Simulation by Multi-Agent Systems.

The simulation by multi-agent systems is adapted to reality, by giving a simple and high quality representation of reality. Multi-Agent Systems are used to find a global solution to a global problem, to model the non-linear, scalable, heterogeneous and complex systems, and to analyse the different interactions between autonomous agents of a global system.

Develop a simulator with AnyLogic platform and preliminary results are presented that will present the drivers, trucks, CO and loading / unloading posts as agents to have a model facilitating this checking.

4.1 AnyLogic Platform

AnyLogic is a simulation platform that includes a graphical modelling language facilitating the models' construction. This platform allows the user performing custom simulation models through its large graphical editor, visual and animated representation (2D and 3D) of the different entities of simulation, and with Java code. Furthermore, it is the first and the only dynamic simulation tool that gathers and combines these three-different modelling-simulation approaches: System Dynamics, Discrete event, and Multi-Agent Systems (MAS).

AnyLogic is the most suitable platform for modelling industrial problems especially the one of the supply chain [3-7].

4.2 Simulation Results

Simulation model provides checking truck before proceeding to loading/unloading operations in order to ensure compliance of trucks to prevent leaks that could create explosion.

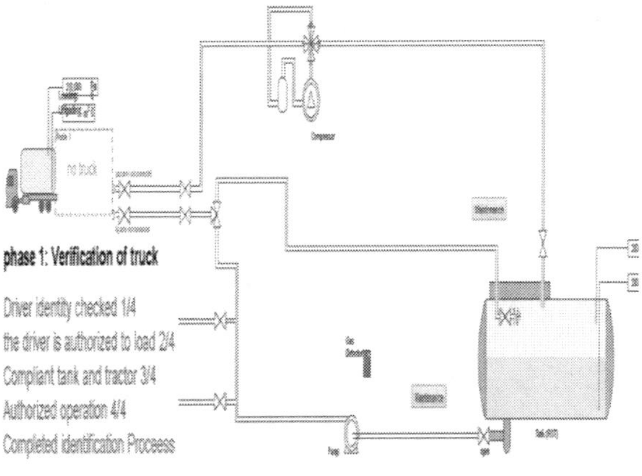

Fig. 4. AnyLogic simulation for checking trucks.

The simulation displays 4 possible cases: Initially when the truck and the tank are complying, **we checked the authorization and driver training** (Fig. 5-8).

If the driver is trained and audited, it is displayed in green:

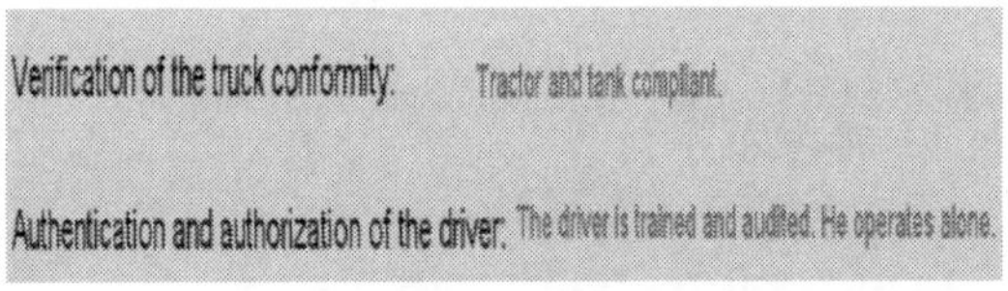

Fig.5. Checking the tuck conformity and authentication of the driver (a).

If the driver is completely formed, it is displayed in blue:

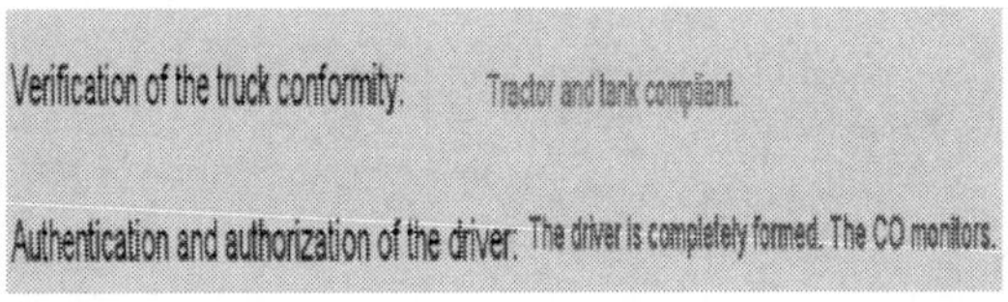

Fig.6. Checking the tuck conformity and authentication of the driver (b).

 DOI: 10.1051/matecconf/201710500004

If the driver is not formed, it is displayed in black (Fig.7):

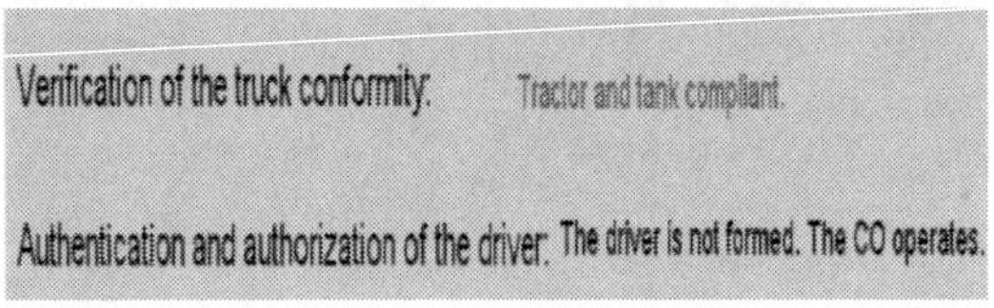

Fig.7. Checking the tuck conformity and authentication of the driver (c).

If the truck or tanks are not compliant, it is displayed in red (Fig. 8):

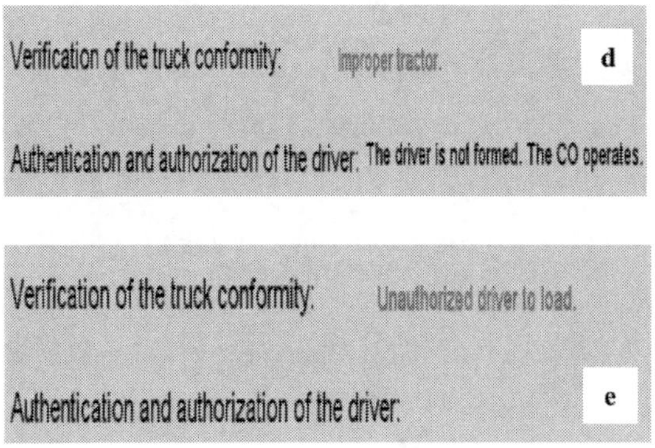

Fig.8. Checking the tuck conformity and authentication of the driver (d, e).

5 Conclusion and Perspectives

LPG Loading /unloading site is considered as a dangerous environment of significant risk, especially during LPG transfer operations. These risks may cause catastrophic dangers such as BLEVE (Boiling Liquid Expanding Vapour Explosion), UVCE (Unconfined Vapour Cloud Explosion), etc.

The modelling-simulation are done using multi-agent systems that present the drivers, trucks, CO and loading / unloading posts as agents in order to have a model facilitating this checking.

Several other improvements can be made to the proposed model and simulator such as ensuring the safety of maintenance by providing a decision support system, for occupational hazard analysis in maintenance tasks, which will allow orienting the actors to the best decisions in order to minimize hazards that may arise. It is a model for risk management for assessing and simulating risk scenarios related to maintenance activities.

References

1. S. Lim. Influence lessons learned from accidents involving LPG storage in the development of guides to good practice abroad. Study Report No. DRA-08-85166-00650B, 04/09/2008 : INERIS for MEEDDAT / SEI / BARPI (2008).
2. M. Gallab, H. Bouloiz, E. Garbolino, M. Tkiouat, M.A. ElKilani, N. Bureau. Risk analysis of maintenance activities in a LPG supply chain with a Multi-Agent approach. J Loss Prevent Proc, **47**, pp 41-56 (2017).
3. M. Gallab, H. Bouloiz, Y. Chater, M. Tkiouat. Towards a simulation Model to ensure the Availability of Machines in Maintenance Activities. Inter. J. of Computer, Electrical, Automation, Control and Information Engineering, **10**, N°7 (2016).
4. C. Almeder, M. Preusser. A hybrid Simulation Optimization approach for supply chains. Proceeding EUROSIM/ SLOSIM, Ljubljana, Slovenia. pp. 1-6 (2007).
5. F. Barahona, M. Ettl, M. Petrik, P. Rimshnick. Agile logistics simulation and optimization for managing disaster responses. Proceedings of the Winter Simulation Conference, 978-1-4799-2076-1/13/$31.00, IEEE, pp. 3340-3351 (2013).
6. S. Kim, S. Mungle, Y.L. Son. An agent-based simulation approach for dual toll pricing of hazardous material transportation. Proceedings of the Winter Simulation Conference. 978-1-4799-2076-1/13/$31.00, IEEE, pp. 2520-2531 (2013).
7. J. Yuan, T. Ponsignon. Towards a SemiConductor Supply Chain Library (SCSC-SIMLIB). Proceedings of the Winter simulation Conference. 978-1-4799-7486-3/14/$31.00, IEEE, pp. 2522-2532 (2014).

Multi-objective optimization approach for air traffic flow management

Rabie Fadil[1], *Badr* Abou El Majd[1], *Hicham* Rahil[2], *Hassan* El Ghazi[3], and *Naima* Kaabouch[4]

[1] *LIMSAD Lab, Faculty of Science Ain Chock - Casablanca, Morocco*
[2] *Air Navigation Department, Mohammed V International Airport,Casablanca, Morocco*
[3] *STRS Lab, National Institute of Posts and Telecommunication, Rabat, Morocco*
[4] *Electrical Engineering Department, University of North Dakota, Grand Forks, United States*

Abstract. Air traffic has long been a generally high growth sector and all forecasts indicate that this trend will continue at a similar pace for the next twenty years. The regular traffic demand growth has led to congestion at airports and in space.
In this paper, we will create in first a probabilistic model which describes the uncertainty of the aircraft's trajectory, and its presence in a sector during a time interval. We define as result, a multi-objective optimization problem whose objective functions are the expected cost of delay and the expected cost of congestion.
Then we use the Non-dominated Sorting Genetic Algorithm (NSGA-II) to solve an instance included 21 flights and 1 sector, and is able to provide a good approximation of the Pareto front.
The decision-making stage was then performed with the aid of data clustering techniques to reduce the size of the Pareto-optimal set and obtain a smaller representation of the multi-objective design space, thereby making it easier for the decision-maker to find satisfactory and meaningful trade-offs, and to select a preferred final design solution.

1 INTRODUCTION

The Operational Research community has studied many variants of the air traffic flow management problem since the beginning of the 90s. Along the years, the models have been refined in order to take into account new operational constraints.

One of the first formulations of air traffic flow management is the ground holding problem, which minimizes the sum of airborne and ground delay costs when the demand for the runways exceeds the allowed capacities. The first work under this formulation has been the Single Airport Ground Holding Problem [1]. After that, [2] has proposed a stochastic and dynamic version of the formulation.

Then, the Multi-Airport Ground Holding Problem was addressed by [3]. This does not take into account the sector capacities,rerouting and speed changes.

The first two limitations were overcome with the model of [4]. Also, this work has the merit to use realistic instances with several thousand flights. To our knowledge, the most comprehensive formulation is the Air Traffic Flow Management Rerouting Problem [5] which integrates all phases of a flight, different costs for ground and air delays, rerouting, continued flights and cancellations. Also, with the same mathematical framework, [6], [7] have formulated the problem in terms of routes instead of nodes.In the same manner, there is also the work of [8] which describes an optimization problem to minimize directly the probability of congestion of the sectors with the concept of chance constraint.

Besides, other techniques were used to solve similar problems. [9] uses stochastic optimization methods for handling sector congestion with take-off delays and rerouting. Constraint programming was also used by [10] and [11]. The former solves the slot allocation problem with sector capacity constraints and the former minimizes an air traffic complexity metric for multiple sectors.

More recently, a multi-objective optimization approach has been used in air traffic control by [11] to minimize an aggregated complexity metric, designed and validated by *Eurocontrol* [1], over sectors. Also, [12] uses the multi-objective to model the trade-off between sector congestion and delays.

2 THE MATHEMATICAL MODEL

In this section, we briefly present the mathematical model wich is based on [13] and [14]. It consists of two submodels, the flight model and the sector model. The former is used to compute the expected cost of delay and the probability of presence in the sectors. The latter takes the probability of presence for each flight

[1] European Organisation for the Safety of Air Navigation, it coordinates and plans air traffic control for all of Europe. This involves working with national authorities, air navigation service providers, civil and military airspace users, airports, and other organisations. Its activities involve all gate-to-gate air navigation service operations.

and compute the expected cost of congestion. First, the flight model is defined as follows:

$$\begin{aligned} p_N(t_N) &= \int_\Omega \dots \int_\Omega p_{1:N}(t_{1:N}) dt_{1:N-1} \\ &= \int_\Omega \dots \int_\Omega \prod_{i=2}^{n} p_{i|i-1}(t_i|t_{i-1}).p_1(t_1)dt_1 \dots dt_{N-1} \\ &= \int_\Omega \dots \underbrace{\left[\int_\Omega p_{3|2}(t_3|t_2) \underbrace{\left[\int_\Omega p_{2|1}(t_2|t_1).p_1(t_1)dt_1 \right]}_{p_2(t_2)} dt_2 \right]}_{p_3(t_3)} \dots dt_{N-1} \end{aligned}$$

Where Ω is the time line, $p_1(t_1)$ is the marginal density function for the flight f to enter the airspace at time t_1 and $p_{i+1|i}(t_{i+1}|t_i)$ is the conditional density function to be on X_{i+1} at time t_{i+1} given it was on X_i at time t_i. Then the probability to be in sector during the time interval $\Delta t = [t_{min}, t_{max}]$ is :

$$P\left(S_{s,f}^{\Delta t}\right) = F_i(t_{max}) - F_j(t_{min}) \quad (1)$$

Where $F_i(t)$ is the cumulative density function denoting the probability that the flight f has flown over X_i by time t.

Thereafter, the probability to have n flights in the sector s during time Δt can be expressed by :

$$P\left(K_s^{\Delta t} = n\right) = \frac{1}{N_s^{\Delta t} + 1} \sum_{l=0}^{N_s^{\Delta t}} exp(-iwln) \, . \prod_{f=1}^{N_s^{\Delta t}} \left[1 - P\left(S_{s,f}^{\Delta t}\right) + P\left(S_{s,f}^{\Delta t}\right) exp(iwl)\right] \quad (2)$$

Where $i = \sqrt{-1}$ and $w = \frac{2\pi}{N_s^{\Delta t}+1}$.

Now, we have all the elements to define our multi-objective optimization problem. Because of the stochastic context, one way to define the cost functions is to use their expected values. So the total expected cost of the delays is defined as follows:

$$\begin{aligned} C_1(\gamma) &= \sum_{f\in\mathcal{F}} E_{\phi_f}\left(T_{n_f}^f; \gamma_{|f}\right) \\ &= \sum_{f\in\mathcal{F}} \left[\int_\Omega (t - A_f)^2 . p_{n_f}^f(t; \gamma_{|f}) dt\right] \end{aligned} \quad (3)$$

Where $\mathcal{F}$ is the set of flights, A_f is the scheduled time of arrival of flight f, n_f the number of waypoints in the flight plan f and p_{n_f} refers to the marginal density function associated to the arrival point $X_{n_f}^f$. $\gamma_{|f}$ denotes the decision vector, the target times of arrival on the different waypoints of the flight f. Indeed, in the optimization context, we are trying to find the vector γ that minimizes the cost functions. For the expected cost of congestion, the function is:

$$C_2(\gamma) = \int_\Omega \sum_{n=C_s+1}^{N_s^{\Delta t}} (n - C_s)^2 . P(K_s^{\Delta t} = n; \gamma) dt \quad (4)$$

Where C_s is the capacity of the sector s and $N_s^{\Delta t}$ is the maximum number of flights that can be inside the sector at the same time with probability higher than 0 and $\overline{\Omega}$ a temporal horizon sufficiently large to encompass the supports of the marginal distributions.

3 EXPERIMENTS

In this section, we describe the experiment on minimizing both the expected cost of delays and the expected cost of congestion with a multiobjective optimization algorithm and a probabilistic model. Then the results on a simple instance are detailed.

The chosen variation operators are shown in the table below:

NSGA-II operators

Encoding: Real
Population size : 1000
Maximal number of evaluations : 30000
Selection : Binary tournament
Crossover : Simulated Binary Crossover , $P_c = 0.9$
Mutation : Polynomial, $P_m = 1/NumberOfVariables$

3.1 Experimental Setting

The chosen instance for this study, implies 21 flights, including 15 departures and 6 arrivals in the Terminal Manoeuvring Area (TMA) surrounding Mohammed V International Airport.

All the results were computed on an Intel® Core™ i5-4200M CPU with 4× 2.5 GHz and 4Go of memory. The model was coded in Java, for NSGA-II and SMPSO we have used the implementation provided by the jMetal framework. We have also used the Apache Commons Mathematics Library for the probabilistic distributions and the computation of the integrals. This library is freely downloaded at : http://commons.apache.org/

3.2 Analysis

Now, we will analyze each step of the methodology. The first step consists in computing the marginal probabilities over the waypoints, as depicted on figure 1, We can see that the value of the modes decreases and the supports of the distributions increase with time. This simply translates the fact that uncertainty on the target time of arrival increases with time. The next step is the computation of the probability for a flight to be in a sector along the time. This is done with (1) and figure 2 shows the result.

The probability increases when the aircraft approaches the sector and once the flight is inside the sector, the probability should not decrease until the exit.

Thereafter, we need to compute the probability that there will be n flights inside the sector at a given time with (2) and we can remark in figure 3 that the probability of congestion decreases with time since the variance of the marginals increases with time.

DOI: 10.1051/matecconf/201710500005

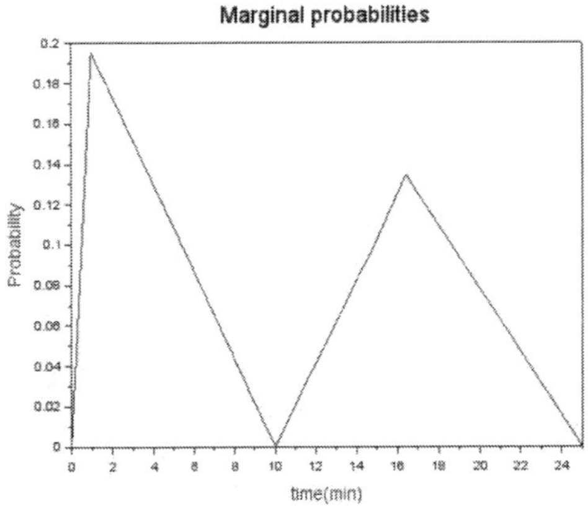

Figure 1: Marginals probability over two waypoints

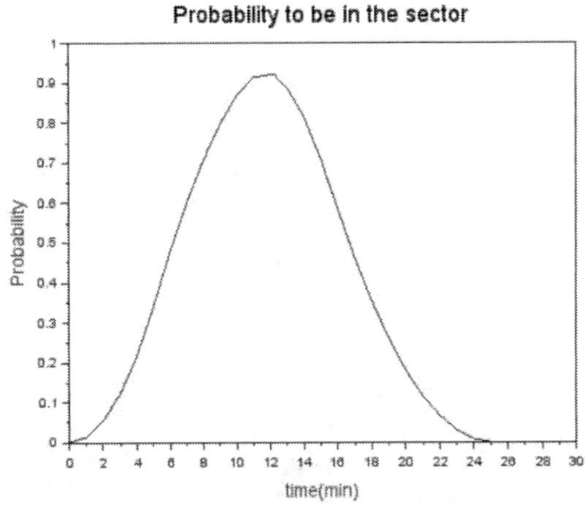

Figure 2: Probability to be in the sector

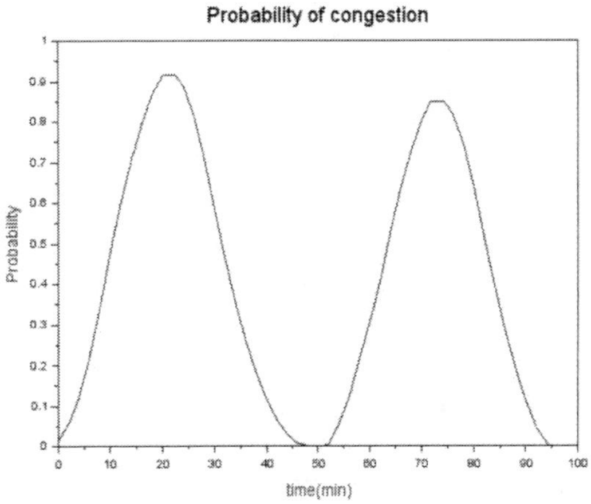

Figure 3: Probability of congestion

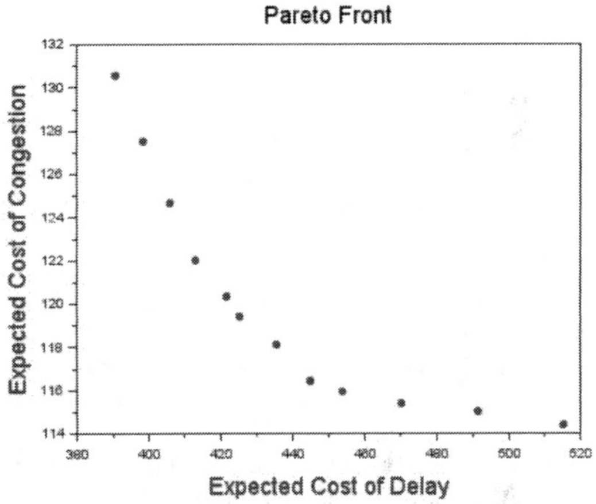

Figure 4: Pareto front obtained with NSGA-II

When all these distributions are known, the probabilistic model can compute the expected cost of delays and expected cost of congestion with (3) and (4) respectively. We can note that minimizing the probability of congestion for every timestamp will effectively minimize the expected cost of congestion.

Finally, when the cost functions are known, one can optimize by given different objectives to the flights. Figure 4 shows the Pareto front for 12 different solutions. Every solution is a complete schedule in the decision space and so, the Pareto front shows the trade-off between minimizing both the delays and the congestion. If the decision maker wants to minimize the chance that delaying the flights congests the sector, he shall select a solution toward the lower right corner of the graph. Otherwise, if she/he believes that the controllers can manage more flights, he can choose a schedule in the top left corner of the graph, resulting in much less delays, at the price of a higher congestion in some sectors.

4 Post-Pareto Analysis

In this section, we attempt to reduce the number of trade-off solutions while considering the stochastic nature of the objective functions.

We will use a standard agglomerative hierarchical clustering technique. We have chosen a hierarchical clustering technique (over a partitioning one, such as k-means) because it allows us to easily adapt the level of granularity of the clusters, and it does not require the number of clusters to be specified in advance.

When carrying out a hierarchical cluster analysis, the process can be represented on a diagram known as a *dendrogram*. This diagram illustrates which clusters have been joined at each stage of the analysis and the distance between clusters at the time of joining.

In our approach, we have generated a dendrogram from the solutions enclosed in the Pareto front as de-

DOI: 10.1051/matecconf/201710500005

picted in figure 4. Figure 5 shows the dendrogram obtained using the *Euclidean distance* as a measure of similarity and the *Ward's method* as a hierarchical clustering procedure.

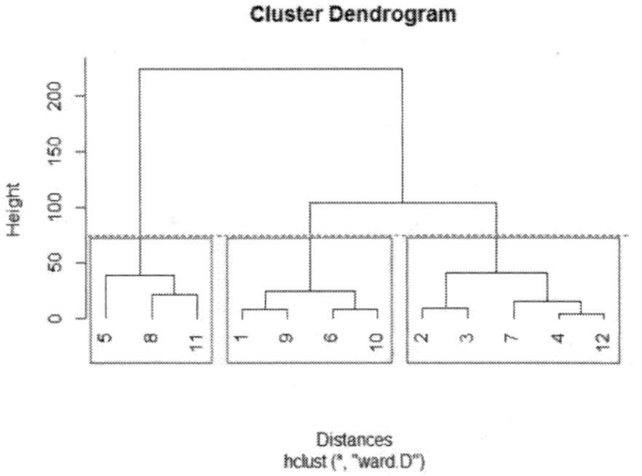

Figure 5: Dendogram

Visual inspection of a dendrogram can give indications about the spread of solutions in the solution. The height of a node represents the distance between its two sub-clusters. Therefore, nodes that are low on the dendrogram form clusters whose elements are closer to each other than nodes that are placed higher up. For example, the cluster 1,9 forms a cluster whose elements are closer to each other than the cluster 5,8,11. The visual inspection of a dendogram can also reveal long branches leading to isolated solutions, which may indicate the presence of "outliers". An outlier in our context is a solution that is radically different from other solutions in the set.

Once the dendrogram has been generated, we need to decide a cut-off at a certain level in order to obtain a set of clusters. For example, with a cut-off value of 60, shown with dotted line on figure 5, we have generated 3 clusters formed by the set of elements under the three branches cut by that line and framed by rectangles.

A recurrent problem that many clustering algorithms encounter is the choice of the number of clusters. Thus, different cluster validity indices have been suggested to address this problem.

The Silhouette [15] and The Elbow method [16] are two of these cluster validity techniques. They are helpful metrics to guide the choice of the best number of clusters.

Figure 6 shows the Scree plot which represents the number of clusters against the total within sum of squares, we can see that the error decreases as the number of clusters gets larger, this is because when the number of clusters increases, they should be smaller, so distortion is also smaller. The idea of the elbow method is to choose the number of clusters at which the SSE decreases abruptly. This produces an "elbow effect" in the graph. In this case, the Elbow method has selected 3 clusters.

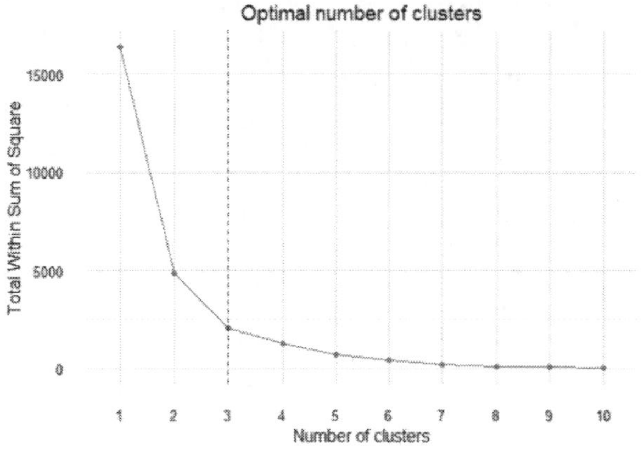

Figure 6: Scree plot

Figure 6 shows the clusters silhouette plot with an average silhouette width of 0.47, which indicates a satisfactory cluster quality. After we have validated the number of clusters, we need to select a representative solution for each cluster. To do this, the solution that is closest to its respective cluster centroid is chosen as a good representative solution. The obtained results of the clustering analysis are depicted in table 1 and figure 8 represents the final solutions.

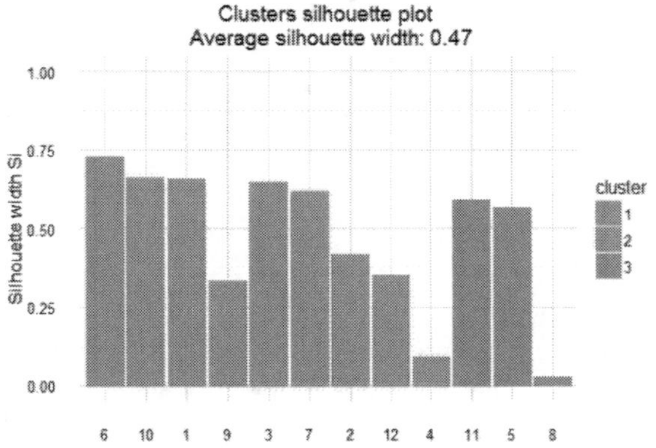

Figure 7: Silhouette plot

5 Conclusion

This paper has presented a probabilistic model to handle the propagation of the uncertainty from the trajectories to the sectors. Then the probabilistic model was used within an optimization algorithm for scheduling all flights at boundaries of the sectors in order to minimize the expected cumulated delays and the expected sector congestion.

Table 1: Final compromise solutions

Cluster	Solutions	Cluster centroid		Representative solutions	
		Expected Cost of Delay	Expected Cost of Congestion	Expected Cost of Delay	Expected Cost of Congestion
cluster1	1,6,9,10	401.97	126.15	398.24	127.48
cluster2	2,3,4,7,12	436.29	118.02	435.51	118.11
cluster3	5,8,11	492.33	114.90	491.5	114.99

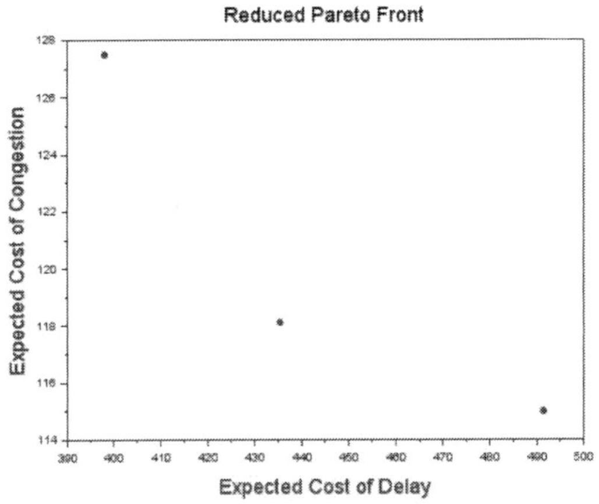

Figure 8: Reduced Pareto Front

NSGA-II was proposed to solve the multi-objective problem, which consists of minimizing both the expected cost of delay and the expected cost of congestion.

Furthermore, in order to illustrate how the theoretical model can be useful in practice, we presented some results on an instance with 21 flights, including 15 departures and 6 arrivals in the Terminal Manoeuvring Area (TMA) surrounding Mohammed V International Airport.

The decision-making stage was then performed with the aid of data clustering techniques to reduce the size of the Pareto-optimal set in order to help the decision-maker to select a preferred final design solution.

References

[1] A. Odoni, " The Management Problem in Air Traffic Control ", *Flow Control of Congested Networks*, pp. 269-288, 1987.

[2] O. Richetta and A. R. Odoni, " Dynamic Solution to the Ground-Holding Problem in Air Traffic Control ", *Transportation Research Part A: Policy and Practice*, vol. 28, no. 3, pp. 167-185, May 1994.

[3] P. Vranas, D. Bertsimas, and A. Odoni, " The Multi-Airport Ground- Holding Problem in Air Traffic Control " ,*Operations Research*, vol. 42, no. 2, pp. 249-261, 1994.

[4] D. Bertsimas and S. S. Patterson, " The Air Traffic Flow Management Problem with Enroute Capacities ", *Operations Research*, vol. 46, no. 3, pp. 406-422, 1998.

[5] D. Bertsimas, G. Lulli, and A. R. Odoni, " An Integer Optimization Approach to Large-Scale Air Traffic Flow Management ", *Operations Research*, vol. 59, no. 1, pp. 211-227, 2011.

[6] A. Agustin, A. Alonso-Ayuso, L. Escudero, and C. Pizarro, " On Air Traffic Flow Management with Rerouting. Part I: Deterministic Case ", *European Journal of Operational Research*, vol. 219, no. 1, pp. 156-166, May 2012.

[7] A. Agustin, A. Alonso-Ayuso, L. Escudero, and C. Pizarro," On Air Traffic Flow Management with Rerouting. Part II: Stochastic Case ", *European Journal of Operational Research*, vol. 219, no. 1, pp. 167-177, May 2012.

[8] G. Clare and A. Richards, " Air Traffic Flow Management under Uncertainty: Application of Chance Constraints ", in *Proceedings of the* 2nd *International Conference on Application and Theory of Automation in Command and Control Systems*, ser. ATACCS '12. Toulouse, France, France: IRIT Press, 2012, pp.20-26.

[9] S. Oussedik, D. Delahaye, and M. Schoenauer, " Air Traffic Management by Stochastic Optimization ", in 2nd *USA/Europe RD Seminar on AirTraffic Management*, 1998, pp. 1-10.

[10] N. Barnier, P. Brisset, and T. Rivière, " Slot Allocation with Constraint Programming: Models and Results ", in 4th *USA/Europe RD Seminar on Air Traffic Management*, 2001.

[11] P. Flener, J. Pearson, M. Ågren, C. Garcia Avello, M. Celiktin, and S. Dissing, " Air-Traffic Complexity Resolution in Multi-Sector Planning using Constraint Programming ", in 7th *USA/Europe RD Seminar on Air Traffic Management*, 2007.

[12] D. Delahaye, S. Oussedik, and S. Puechmorel, "Airspace Congestion Smoothing by Multi-Objective Genetic Algorithm ", in *Proceedings of the 2005 ACM symposium on Applied computing*, ser. SAC '05. New York,ACM, 2005, pp. 907-912.

[13] G. Marceau Caron, P. Savéant and M. Schoenauer, "Multiobjective tactical planning under uncertainty for air traffic flow and capacity management ",*IEEE Congress On Evolutionary Computation. IEEE Computational Intelligence Society*,2013.

[14] A. Hadjaz, G. Marceau Caron, P. Savéant and M. Schoenauer, "Increasing Air Trafc: What is the Problem ? ",*The Second SESAR Innovation Days* ,2012.

[15] P. J. Rousseeuw,(1987). " Silhouettes: a graphical aid to the interpretation and validation of cluster analysis ". *Journal of Computational and Applied Mathematics.* Vol.20, pp.53-65.

[16] Robert L. Thorndike (December 1953). " Who Belongs in the Family ? ". Psychometrika .18 (4): 267-276. doi:10.1007/BF02289263.

MATEC Web of Conferences **105**, 00006 (2017) DOI: 10.1051/matecconf/201710500006
IWTSCE'16

Impact of several parameters on accuracy RFID speed measurement system

Laila Abouzaid[1], Ahmed Errami[1], Othmane Benhmammouch[1,2]

[1]*RTSE Hassan II University, ENSEM-Casablanca, Morocco*
[2]*International University of Casablanca, Morocco*

Abstract. Radio Frequency Identification (RFID) is a rapidly growing automatic wireless data collection technology, this technology is applied in a variety of applications, spanning from Supply chain, payments, Race-timing, to transportation. Recently, this technology is becoming increasingly present in some new fields. In this paper we focus on the RFID speed monitoring system. In this effect, a method for verification of speed accuracy is proposed. The proposed method can illustrate the error introduced by each parameter contribute to calculate speed of vehicle using RFID speed system.

1 Introduction

*"**Traffic accidents are predicted to become the second leading cause of the death in developing countries by the years 2020**"* [1]

At Over 1, 2 million human deaths every year on the world's, due to road traffic crashes, and between 20 and 50 million sustain non-fatal injuries. Unfortunately, for almost all countries, in three out of four cases the deceased persons affect young adults who involve excessively in economically productive activity [2].
There are numerous contributory causes of road traffic accidents, which fall into three categories: (1) Meteorological factors caused by the snow, rain or sleet, can provide poor visibility. (2) Technical factors include generally vehicle status and road infrastructures. (3) Human factors, is concerned the both drivers and road users; which appear in all other causes and represent 90% of fatal crashes. Alcohol, drugs, fatigue, the use of Mobile Phone, not stopping at red lights and breaking the speed limit, are contributory factors that lead to road accidents. However, it is estimated that over 40% of fatal crashes are caused by excessive or inappropriate speed [1,3,4].
The road safety strategy, in all the countries propose a plan to achieve a reduction in road deaths and serious injuries, the main action strategy to reduce a numbers of accidents linked to speeding, setting out the generalization of speedometers, in the all roads to control the speed with appropriate sanctions [5]. For that reason, this paper will focus on RFID speed measurement systems in order to examine the factors that influence the precision of the system cited above.
The remainder of this paper is structured as follows: Section 2, gives an overview of widespread speedometers, section 3, describes the RFID speed monitoring system, followed by a problem of precision provided by this system, section 4 explains the factors that contribute in this problem and the influence of these parameters on average speed is presented in section 5.

2 Overview of Speed Measurement Systems

The ubiquitous systems of vehicle's speed on the road are mainly split in two categories: Speed detection with Doppler Effect and laser speed detectors (LIDAR). The first type operates according to Doppler shift which based on the change of frequency of an electromagnetic or sound waves emitted by a fixed source and reflected by the moving objects of interest, this frequency is proportional to the speed of the moving object. On the other hand the principle of the second type requires the determination of the distance between the radar and target; is given by the measurement of the delay between the instant of transmission of a pulse and the instant of detection of the reflected pulse, knowing the speed of light. The variation of the distance obtained by the emission of a periodic pulse burst used to deduce target speed [6,7].

DOI: 10.1051/matecconf/201710500006

However, the two techniques mentioned earlier plagued by the difficulty to identify vehicles exceeding the speed limit.

Various solutions have been proposed over the years to solve the problem posed, thus afford to measure the vehicle speed. One possible solution based on the use of a coupling between Doppler radar or the laser detector with imaging device configured to capture at least one image or multiple images. Evidently, the acquired scene image offers the license plate number of the vehicle and/or identification of the driver face; in case of infringement, these data are then used to identify the driver and vehicle. However, the image taken in bad weather conditions or the presence of obstacles may increase the risk to recognize vehicles in cases of violations.

There are well-known solutions, resorts to the use of imaging devices which can be generate traffic video and/or capture images whose purpose is to determine and identify vehicles exceeding the speed limit. Unfortunately, the collection of information by Camera video-based speed enforcement system lead to a huge storage of the acquired data, which requires a complicated procedure for pre-process to recognize the license plate number and other information is needed [8,9].

The aforementioned drawbacks are substantially eliminated by the proposed system, in which the speed measuring of a vehicle and identification of this target are performed using the RFID technology and the Doppler Effect. The documents [10] ;exhibit this system that has as principle the simultaneous combination of the acquisition of information relating to the vehicle first determining the speed of the latter through a tag attached to the vehicle. The calculation of the speed is done by using the frequency shift by Doppler Effect extract, from the signals transmitted and received between the RFID reader and the RFID tag. According to the above technology, the work [11] present a new vehicle speed monitoring system, based on an RFID communication.

3 RFID speed measurement system

3.1 Description

The RFID speed monitoring system, as mentioned above consists of two RFID readers placed at determined distance D, able to get the current information about the vehicle embedded in the transponder and to memorize time of a vehicle passing through the reader interrogation zone. Knowing the times Δt and distance travelling between two points we can calculate the speed of each vehicle travelling on road using the basic formula for average velocity (1).

$$V = D/_{\Delta t} \quad (1)$$

The model (figure 1) represent side-looking configuration, that have an illuminative direction perpendicular to traffic, this configuration is the most popular approach for current ITS (intelligent transportation systems) applications which can cover all the lanes [12].

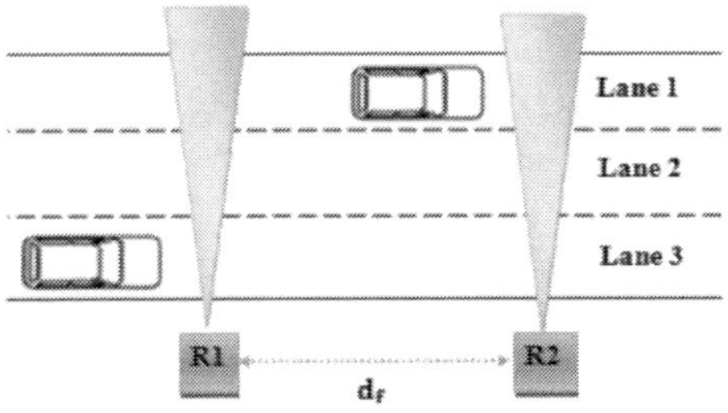

Figure 1: side-looking RFID radar systems.

3.2 Error analysis

RFID speed monitoring system has some desirable features, such as low cost, not requiring line of sight between readers and labels to detect the target vehicle. However, such devices suffer from the limited accuracy of average speed especially in the event where the roadway is composed of several lanes and in the presence of multiple vehicles. To insure a well-functioning, it's primordial to study all the system aspects.

One of the main disadvantages is that a reader's log the times which are not adapted to a transmission positions. This notion is also extremely important, because it can easy miss significant changes on travelled distance D, which can provide error in speed measurement. In other words the speed error provided is caused by a shift position between the reference axis and the line connection to the tag (Δd_1 , Δd_2); weather it is for the first reader R1, or the second reader R2 (Figure 2).

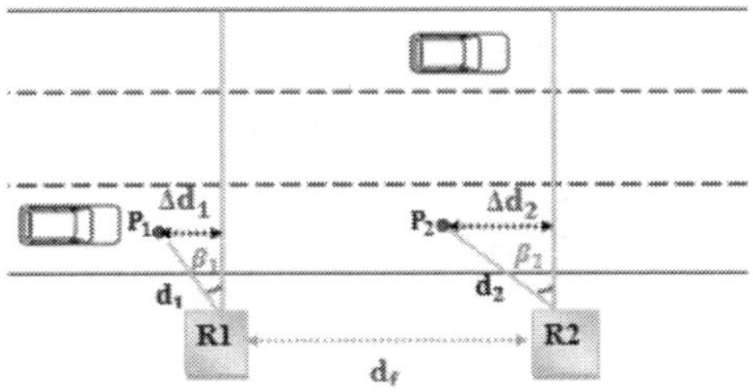

Figure 2: Components of RFID speed measurement system.

This paper aims to determine the impact of all parameters contributing to the RFID speed measurement system and their impact on speed accuracy.

4 THE SPEED-ACCURACY

The first stage of this work will be the determination of all factors needed to calculate speed. We'll start by expressing the travelled distance D by the target with reference to figure (1):

$$D = d_f + \Delta d_1 - \Delta d_2 \quad (2)$$

Where d_f is the distance between two readers, and Δd_1, Δd_2 are the maximum error to detecte the position of transponder by the first and second reader respectively,with :

$$\Delta d_1 = \text{Sin}(\beta_1) * d_1 \quad (3)$$
$$\Delta d_2 = \text{Sin}(\beta_2) * d_2 \quad (4)$$

Where d_1, d_2 are the distances between the readers and tag attached to vehicles, β_1, β_2 are the angle between the reference axis and the line connection with the tag, we can express :

$$\beta_2 = \beta_1 \pm \mathcal{E} \quad (5)$$

Where $\mathcal{E}$ the variation of detection angle by the two readers. According to Friis Transmission Equation we can obtain [13] [14]:

$$(d_1, d_2) = \frac{\lambda}{(4\pi)}\sqrt{P_{EIRP} G^{Tag}\, \tau_{tag}\, \chi \frac{1}{P_R^{Chip}}} \quad (6)$$

Where P_{EIRP} The maximum power output available for RFID applications, G^{Tag} is the gain of the tag antenna respectively , τ_{tag} is the reflection coefficient between the tag antenna and the chip, χ is the polarization factor between reader antenna and tag antenna, λ wavelength and P_R^{Chip} is received power at the chip.

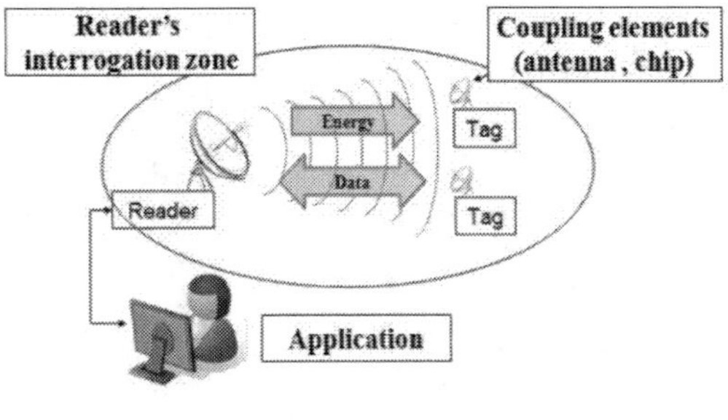

Figure 3: RFID Communication.

Figure 3 illustrates the standard communication between an RFID tag and an RFID reader. The reader transmits an interrogation signal which energizes and activates the RFID tag. Then the tag detects this signal, it responds and is identified [15]. For simplification only a single transponder considered present in the interrogation zone, however if several tags are simultaneously in the area covered by the reader, it will be necessary to integrate collision avoidance algorithms to ensure reading all these transponders [16]. The expression of the Δd_1 and Δd_2 is given in Equation (7) and (8) according to equation (6):

$$\Delta d_1 = \text{Sin}(\beta_1)\frac{\lambda}{(4\pi)}\sqrt{(P_{EIRP}\, G^{Tag}\tau_{tag}\chi \frac{1}{P_R^{Chip}}} \quad (7)$$

$$\Delta d_2 = \text{Sin}(\beta_1 + \mathcal{E})\frac{\lambda}{(4\pi)}\sqrt{(P_{EIRP}\, G^{Tag}\tau_{tag}\chi \frac{1}{P_R^{Chip}}} \quad (8)$$

$$V = \frac{d_f + \Delta d_1 - \Delta d_2}{T} \quad (9)$$

$$V = \frac{d_f + \left(\frac{\lambda}{(4\pi)}\sqrt{(P_{EIRP}\, G^{Tag}\tau_{tag}\chi \frac{1}{P_R^{Chip}}}\right) * \left(\text{Sin}(\beta_1) - \text{Sin}(\beta_1 + \mathcal{E})\right)}{T} \quad (10)$$

5 RESULTS AND DISCUSSION

Furthermore, we study the influence of system parameters on the measurement error generated by speed computing. The first parameter to be analysing is d_f, in order to determine the optimum choice for this factor. Speed Error curve is plotted according to equation (12) by substituting different Δd_1 values and using the values listed below in Table 1.

$$\text{Error speed } (\%) = \frac{\text{Speed}_{\text{Approah}} - \text{Speed}_{\text{sensed}}}{\text{Speed}_{\text{sensed}}} * 100 \quad (11)$$

$$\text{Error speed}(\%) = \left|\frac{\left({}^{d_f}/_{T}\right) - \left({}^{d_f + \Delta d_1 - \Delta d_2}/_{T}\right)}{\left({}^{d_f + \Delta d_1 - \Delta d_2}/_{T}\right)}\right| * 100 \quad (12)$$

Where $\text{Speed}_{\text{Approah}}$ is the speed calculated using D traveled by target equal to distance a fixed distance of readers ($D = d_f$), then $\text{Speed}_{\text{sensed}}$ is calculated by using the equation (10).

Table 1: Parameters used in curves.

Parameters	Values
T(s)	1
Δd_2 (m)	4

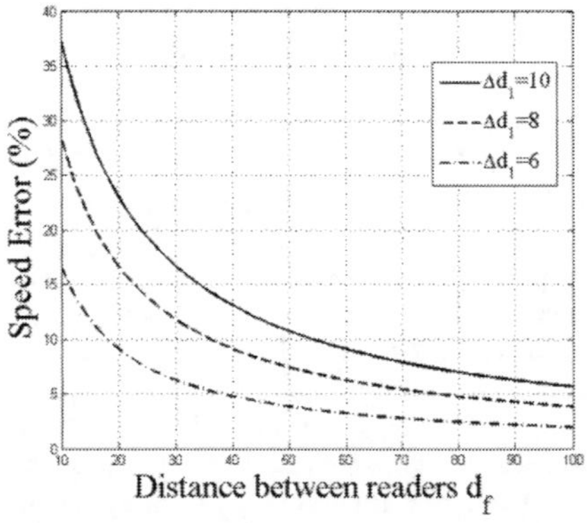

Figure 4: Speed error as function of distance between readers d_f for differentshift position Δd_1(m).

DOI: 10.1051/matecconf/201710500006

As shown, the increasing distance between two readers, will decrease the speed error for different values of variation in detection Δd_1.However, the placement of the two readers at far distance may lead to erroneous results, due to changing lanes or decrease the speed between two consecutive readers. To avoid this type of error we assume that the minimum distance between two readers can be expressed by:

$$d_f = 2 * R * \sin(\Theta) + s \quad (13)$$

Where the Θ is the half aperture angle of reader's antenna, R is maximal read range (14) and s is the security distance to avoid the interference between the reader's interrogations zone (figure 5) which can we take10% of R.

$$R = \frac{\lambda}{(4\pi)} \sqrt{P_{EIRP}\, G^{Tag}\, \tau_{tag}\, \chi\, \frac{1}{P^{Chip}_{Rmin}}} \quad (14)$$

$$d_f = \frac{\lambda}{(4\pi)} \sqrt{P_{EIRP} G^{Tag} \tau_{tag} \chi \frac{1}{P^{Chip}_{Rmin}}} * (2 * \sin(\Theta) + 0.1) \quad (15)$$

With P^{Chip}_{Rmin} is the threshold power of a transponder. In case where the two readers have the same characteristics and the width of the road varies from 14 to 16 meters, the distance d_f won't exceed the 32 meters. Unfortunately, this distance become $d_f = 32$, according to the figure (4), this value tend to provide significant error.

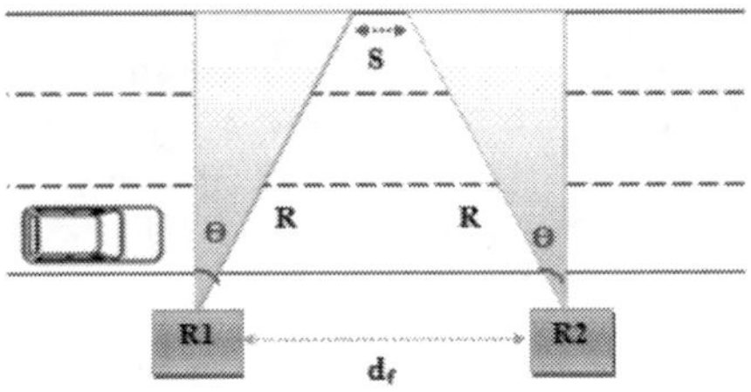

Figure 5: d_f parameters.

According to the equation (10) we have noted that the speed depends on other parameters of the transponder. In this effect we sought to determine the impact of each one (12). Using table 2, we obtain the impact of some parameters in the speed measurement

Table 2: Parameters Used in Simulation.

Parameters	Values
Equivalent Isotropic Radiated Power	500 mW ERP
Polarization factor	-3 dB
Tag power sensitivity	-35 dBm
Θ	40°
β_1	15°
T(s)	1

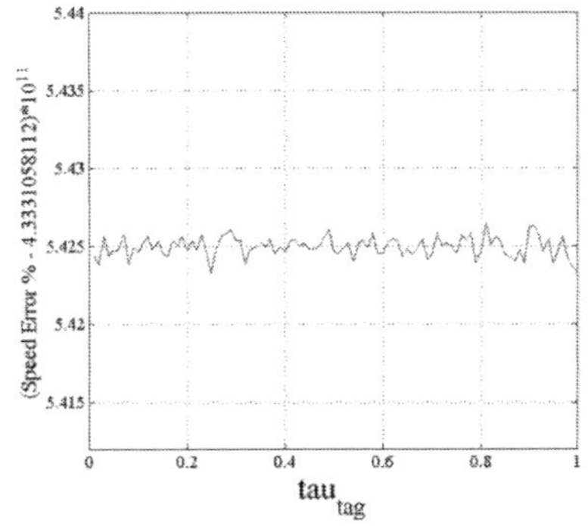

Figure 6: relative speed error as a function of the reflexion coefficient. Other parameters: 867.2 MHz, G^{Tag}=2dBi, P_R^{Chip}=-25dBm ,Ɛ=10°.

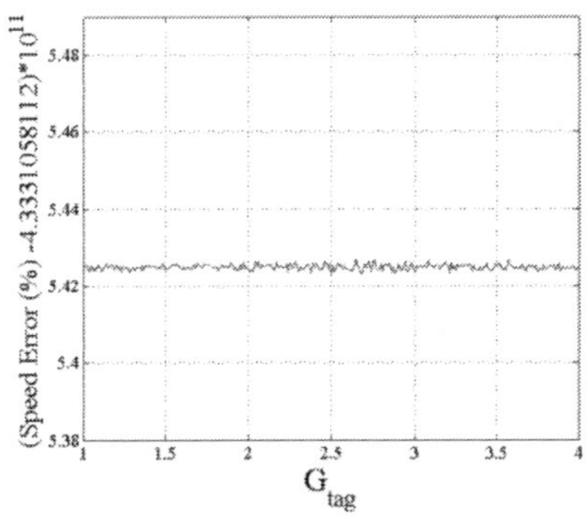

Figure 7 : Speed error as a function of the gain (dBi). Other parameters: 867.2MHz, τ_{tag}=0.8, P_R^{Chip}= -25dBm,Ɛ=10°.

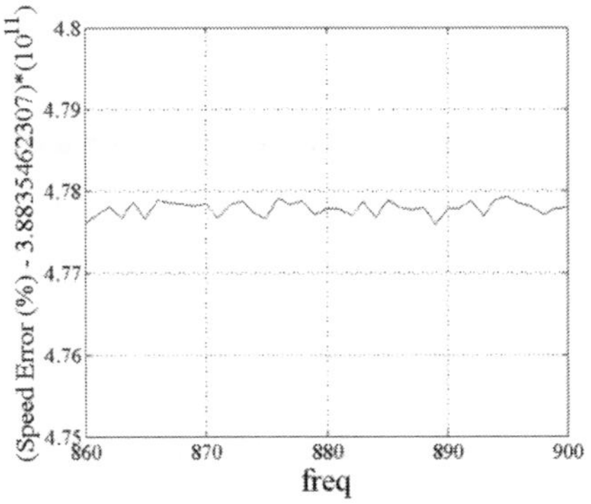

Figure 8: Speed error as a function of frequency (MHz). Other parameters, G^{Tag}= 2dBi, τ_{tag}=0.8, P_R^{Chip}= -25dBm, Ɛ=10°.

DOI: 10.1051/matecconf/201710500006

The figure (6,7and 8) shows a slight error occurs by these parameters (gain G^{Tag}, reflexion coefficient τ_{tag}, frequency) Generally, we can note that the error effecting by these parameters can be imperceptible and can be neglected.

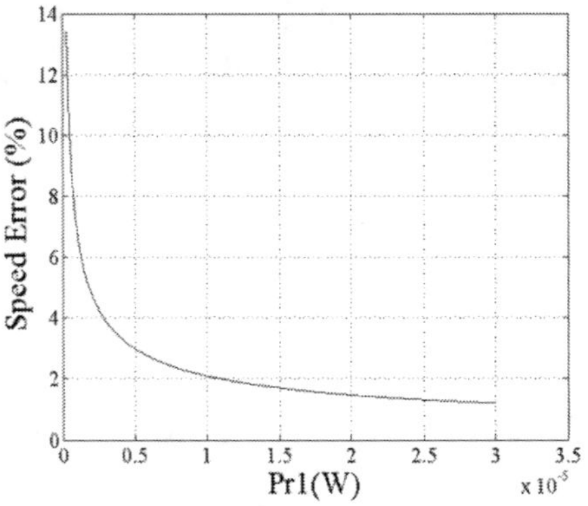

Figure 9: Speed error as a function received power at the chip. Where 867.2 MHz, G^{Tag}=2dBi, τ_{tag}=0.8, Ɛ=10°

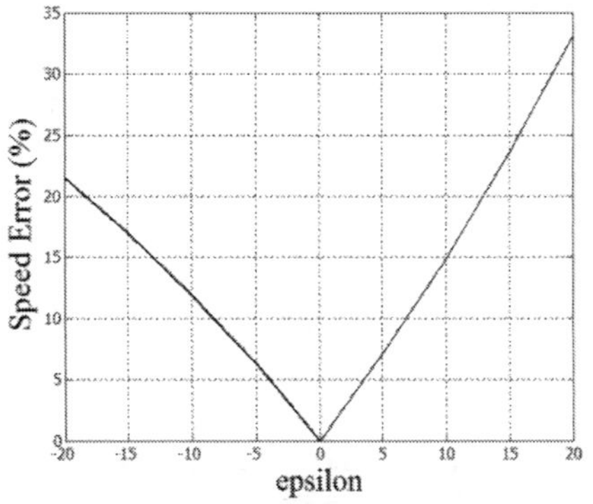

Figure 10 :Speed error as a function of Ɛ , where867.2 MHz, G^{Tag}=2dBi, τ_{tag}=0.8, P_R^{Chip}=-25dBm.

The figure (9) presents the impact of received power at the chip on speed error. The error grows with decreasing received power at the chip (P_R^{Chip}) due to the fact that the parameter mentioned is the main factor which control the distance between reader and target. Hence, the error grows when the target is so far to the reader with the narrow opening angle or wide opening angle. So, we can say that the power received by the chip is inversely proportional to the speed error. The figure 10 reveals with the increasing Ɛ the variation of detection angle the speed error increase.

However the error speed provided by the RFID Speed measurement system is related in the first hand to the parameters Ɛ, and the second hand to received power at the chip P_R^{Chip}.So, in order to compute and have trusted information regarding the speed, we intend to normalize the Ɛ value such a way that the margin of error or tolerance must not exceed 5% comparatively to speedometers existing.

6 Conclusion & future works

This paper gives in the first hand a comparative overview of vehicle speed measurement systems, and then we focus especially on new RFID speed measurement system. In the second hand, we presented the method used to calculate average speed by this system by referring to a model of communication RFID. In fact this work is principally concerned with the parameters that influencing on vehicle speed measurement using RFID technology.

The power received at the chip and the variation of detection angle are the key factors affecting the accuracy of the average speed.

As a future work we are working towards to reducing the speed error by normalize the variation of angle detection, and to find the minimal opening angle of RFID reader antenna, to concentrate the power toward each tags , while insure the read of tags embedded in vehicle driving at high speed. Afterwards we intend to find a model of tag specific applicable RFID speed measurement system.

References

1. B. Bakama Nume , *Road Traffic Accidents in Uganda: Epidemiological and Transport Policy Implications*(African Social Science Review, 2006).
2. M. Peden, R. Scurfield, D. Sleet, Di. Mohan, A. Hyder, E. Jarawan and C. Mathers, *World report on road traffic injury prevention* (World Health Organization Geneva 2004).
3. O. Bagadadi and A. Varhelyi, *An indicator of accident proneness* (Accident Analysis and Prevention, vol. 43, pp. 1359-136 , 2011).
4. Institute of safety road , *The relation between speed and crashes*(SWOV Fact sheet, April 2012).
5. T. Litman, S. Fitzroy, *Evaluating Mobility Management Traffic Safety Impacts* (Victoria Transport Policy Institute, 12 September 2016).
6. S. Park, T. Kim, S. Kang, and K. Heon Koo, *A Novel Signal Processing Technique for Vehicle Detection Radar*, (EEE MIT-S Digest, 2003).
7. Mphippen, *Difference Between Radar and Lidar Explained*, (IPTM Special Problems in

Traffic Crash Reconstruction Conference , May 2013).

8. M. Ayub Khan, S. Ziauddin, I. Haq, N. Mehmood, M. Hussain, S. Hassan, M. Tayyab, *Apparatus and method for automatic license plate recognition and traffic surveillance*,US20150248595 A1,3 sept. 2015
9. W. Wencheng, Edgar A. Bernal, Robert P. Loce, Thomas F. Wade, Abu Islam, *Single camera video-based speed enforcement system with a secondary auxiliary RGB traffic camera*, US9183746 B2,nov 2015.
10. K. Park, T. Kang, S. Hyun, S. Kang, B. Choi, *Method for measuring speed of vehicle using RFID, RFID reader for measuring speed of vehicle, and system for collecting vehicle information using the same*,US8274371 B2,25 sept. 2012.
11. A. Al-Gindy, A. Tawfik, L. Sakhi , *RFID Speed monitoring system* , (Sixth International Conference on Developments in eSystems Engineering, 2013).
12. S. Jeng, W. Chieng, and L. Hsiang-Pin, *Estimating Speed Using a Side-Looking Single-Radar Vehicle Detector* (IEEE transactions on intelligent transportation systems, vol. 15, no. 2, april 2014).
13. M. Dhaouadi , *Conception et optimisation des antennes RFID UHF en vue d'améliorer la fiabilité des systèmes RFID* (2014).
14. P. Jankowski-Mihulowicz, M. Weglarsk, *Determination of Passive and Semi-Passive Chip Parameters Required for Synthesis of Interrogation Zone in UHF RFID Systems*, (ELEKTRONIKAIR ELEKTROTECHNIKA, ISSN 1392-1215, VOL. 20, NO. 9, 2014).
15. H. Peter Cole, Zhonghao Hu and Y. Wang, *Operating Range Evaluation of RFID Systems* (The University of Adelaide, Australia).
16. D. Klair, K. Chin, R. Raad, *A survey and tutorial of RFID anti-collision protocols* (IEEE Communications Surveys & Tutorials, vol.12, (3) pp. 400-421, 2010).

Maturity of the Warehousing function in Moroccan companies: a case study

Mohamed RAZIK[1,*], Bouchaib RADI[2], Chafik OKAR[3]

[1] *Faculty of Sciences and Techniques, University Hassan 1er, Settat, 577, Route de Casablanca, Morocco*
[2] *Faculty of Sciences and Techniques, University Hassan 1er, Settat, 577, Route de Casablanca, Morocco*
[3] *School of Technology, University Hassan 1er, 3 Berrechid, 218 Quartier Taqadom Passage d'Alger, Morocco*

Abstract. In today's competitive market environment, companies are continuously forced to improve their warehousing operations. To meet high performance goals, tools must be developed to improve efficiency in every aspect of warehousing function. In this paper, we propose a new approach to improve the performance of the warehousing function based on Maturity model, which may better identify, explain, assess and improve this critical function in a supply chain. The aim of this article is to validate the model developed by a case study in a Moroccan company by demonstrating its capacity of assessing the maturity of warehousing function and whether it can develop an improvement roadmap.

1. INTRODUCTION

Warehousing continue to play an important role in modern supply chains [1]. In today's competitive market environment, companies are continuously forced to improve their warehousing operations [2]. To meet high performance goals of warehousing, a way needs to be found to eliminate any waste from the warehouse, to streamline its operations, and improve efficiency in every aspect of warehousing activities [3].

Many works in various industrial sectors have shown the benefits of using maturity models and the correlation between the improvement of project maturity and project performance (Ibbs & al 2004 [4], Cooke-davies 2004 [5], Qureshi & al 2009 [6], Pöppelbuk & Röglinger 2011 [7]). They promote organizational learning as well as enabling efficient and effective assessment of the performance management practices of the organizations [8]. Their use for self-assessment should result in growth in the maturity levels of performance management practices that should lead to improved levels of performance [8].

However, despite the existence of maturity models for the supply chain (McCormack & al 2008 [9], Reyes & al 2010 [10]) and performance measurement systems (Wettstein and Kueng 2002 [11]), there is no specific maturity model for the warehousing function. This is why we propose in this article a new approach to improve the performance of the warehousing function based on Maturity model, which may better identify, explain, assess and improve this critical function in a supply chain.

In this paper, we will describe a maturity model for the warehousing function in Moroccan companies developed using the concept of critical success factors (CSFs). In order to validate and to demonstrate the effectiveness of the proposed Model, we will present a case study in a Moroccan company.

2. THE WAREHOUSING FUNCTION

2.1 The importance of the warehousing function in modern supply chains

Supply chain management (SCM) has been a major component of competitive strategy to enhance organizational productivity and profitability [12]. They are characterized by many activities and actors that generally pursue conflicting objectives [13]. The correct organization of all the logistic processes and activities that take place within a supply chain node could have a remarkable impact on both processes upstream and

* m.razik@uhp.ac.ma

downstream the supply chain and on supply chain node internal costs as well [14].

Warehousing have an important role in modern supply chains [15]. A survey of logistics costs in Europe identified the cost of warehousing as being 24% of total logistics costs [16]. A similar study in the USA, found warehousing costs at 22% being close to the European figure [17].

As well as being significant in cost terms, warehousing is also important in terms of customer service, it being critical to the success or failure of many supply chains [18]. According to Werling, the role of the warehouse has changed dramatically over the past 30 years, as customer and vendor compliance issues have surfaced and a greater emphasis has been placed on supply chain visibility and customer satisfaction [19].

Warehousing plays a critical role in ensuring high levels of customer service and overall supply chain performance [20]. For logistics managers, the warehouse is at the center of reflections and challenges, as it has become a real factor in optimizing the supply chain. Warehouse management may strongly affect supply chain performances [14]. While there are widely accepted benchmarks for individual warehouse functions such as order picking, little is known about the overall technical efficiency of warehouses [21]. This creates therefore a need for development of tools for assessment the warehousing function for better its performance improvement.

2.2 The importance of the warehousing function in Moroccan companies

In Morocco, the awareness of the importance of improving the performance of critical business processes is currently greater. Indeed, The Moroccan Company operates in an environment characterized by a competitive offer increasingly strong requirement for competitiveness increasingly acute and expectations of customers increasingly high [22]. To address these issues and to improve performance and competitiveness, all processes must be improved.

According to (Naciri, 2010), the organization of the supply chain of Moroccan companies generates costs that penalize competitiveness [23]. The links in this chain must be therefore mastered and optimized to gain in overall performance. According to a report by the World Bank and the Ministry of Equipment and Transport of the Kingdom of Morocco in 2006, it was reported that the poor organization of warehousing is one of the obstacles to the development of modern logistics in Morocco [24]. Furthermore and according to a study by the Centre for Transportation Studies for the Western Mediterranean, the level of warehousing services in Morocco is relatively low and the technical conditions under which it occurs are lower than the standards [25].

Particular intention must therefore be given to the performance of the warehousing in Moroccan companies, it is not the construction of warehouses that is difficult, but it is the management of these units that can be problematic because it requires needs in know - how Logistics [23].

On the other hand, Morocco lags far behind in terms of logistics platforms [26]. this delay will be caught up with the national strategy of logistics development by creating a network of logistics centers in different morrocan cities [27]. Once these projects are completed, it will be essential to master the warehousing function for better management and operation of these platforms.

2.3 Critical Success Factors (CSFs) for warehousing performance improvement

Across the supply chains, warehousing is an important element of activity in the distribution of goods, from raw materials and work in progress through to finished products. It is not a 'Stand-alone' element of activity and it must not be a weak link in the whole supply chain network [28].

As previously reported, warehousing is costly for enterprises, either in terms of the facilities and equipments required or in terms of human and management resources. Otherwise, the underperformance will risk the achievement of required customer service levels and the maintenance of stock integrity, and result in unnecessarily high costs [28].

Through a previous study [29], we have been determined Critical Success Factors (CSFs) for warehousing performance improvement. We have been classified them into four categories reflecting the four components of the warehousing function defined in our previous study. The table 1 summarizes them:

Table 1. Critical Success Factors (CSFs) for warehousing performance improvement [29]

Critical Success Factors for the component "Design of the warehouse"	▪ Definition of an optimal location of the Warehouse ▪ Definition of an optimal design of the warehouse ▪ Definition of functional areas in the Warehouse ▪ Definition of warehouse storage capacity ▪ Definition of the developments of the warehouse storage capacity in the future

Critical Success Factors for the component "Means for the warehouse"	▪ Selection of the appropriate storage systems ▪ Selection of the appropriate handling equipment ▪ Definition of the necessary and efficient warehouse staff ▪ Use of ICTs
Critical Success Factors for the component "Operations in the warehouse"	▪ Definition of a rule for the allocation of products to the functional areas ▪ Definition of a rule for the allocation of products to the storage areas ▪ Definition of a policy of picking ▪ Definition of a policy of delivery ▪ Setting a policy for the return processing ▪ Optimization of internal logistics for operations in the warehouse
Critical Success Factors for the component "Management of the warehouse"	▪ Definition of inventory management policy ▪ Definition of the staff allocation procedure to different missions ▪ Definition of the quality control procedures in the warehouse ▪ Definition of the security control procedures in the warehouse ▪ Formalization of all working procedures ▪ Use of a warehouse management information system

3. MATURITY MODEL FOR THE WAREHOUSING FUNCTION IN MOROCCAN COMPANIES

3.1 The concept of maturity models

Firms look increasingly to the development of new tools and models to better manage their projects for ensuring the fulfillment of objectives, for increasing the probability of success, and for ensuring the overall performance of the company. In this context, "more and more organizations are relying on process improvement to improve their probability of success" [30]. The maturity models address the need of process improvement [31].

Maturity describes a "state of being complete, perfect or ready." [32]. To reach a desired state of maturity, an evolutionary transformation path from an initial to a target stage needs to be progressed [33].

The concept of the process maturity was born in the Total Quality Management (TQM) movement and it was widely adopted in "Capability Maturity Model" for software organizations [34]. Then this concept migrated to organizational process and project management [35].

Maturity models act as a structured framework for the revolutionary growth of performance within projects or programmes in organizations. They help in assessing the organizational strengths and weaknesses and in bringing the organization to the next level of maturity or achievement in accordance to the goal to be achieved [36]. They provide means of identifying some crucial steps to be taken, the tasks that are necessary to accomplish and the sequence of events needed to realize significant and quantifiable results [37].

The basic concept of all models is based on the fact that things change over time and that most of these changes can be predicted and regulated [38].

3.2 Development of a maturity model for the warehousing function in Moroccan companies

Given the need for a model that assesses the maturity of the warehousing function, and given the absence in the literature of a specific maturity model to this function, we develop in a previous study [39] a preliminary version of a maturity model for warehousing function in Moroccan companies based on the concept of CSFs. The proposed model will be used to identify and exploit the strengths and weaknesses of the warehousing function in Moroccan companies. It will serve as an assessment tool that will improve warehousing performance and consequently that of the supply chain and companies in general.

The structure of the proposed maturity model is built upon the following three dimensions:

1. Maturity level dimension
2. The four main components of the warehousing function;
3. Critical Success Factors (CSFs) for warehousing performance improvement in Moroccan companies.

The first dimension is related to maturity levels, most maturity models measure the maturity through the achievement of some levels range in general from Level 1 to Level 5. In our case, we take a scale of 1 to 3 since we propose a new model that is not developed in the literature, and to facilitate its implementation by Moroccan companies:

- Level 1(initial): there is no process area and process is chaotic;
- Level 2 (defined) : is the level where warehousing function processes are documented, standardized, and integrated into a standard implementation process for the organization and;
- Level 3 (managed): warehousing function process and activities are controlled and managed based on quantitative models and tools.

As the performance of the warehousing depends on components related to CSFs presented above, the second dimension of our model will be the four main components of the warehousing function:

- Component 1 : Design of the warehouse,
- Component 2 : Resources for the warehouse,
- Component 3 : Operations in the warehouse
- Component 4 : Management of the warehouse.

The third dimension will be the CSFs for warehousing performance improvement in Moroccan companies.

The calculation of maturity level is as follows:

- The maturity level of the CSF j for component i :
  - ➢ M CSF ij
- The maturity level of the component i :
  - ➢ M component i = min (M CSF ij)
    j = 1 … .m
- The maturity level of the Warehousing function :
  - ➢ M Warehousing = min (M component i)
    i = 1… n

Based on this principle, we can locally assess the level of maturity and obtain a score for a given CSF, and consequently to a particular component of the warehousing function. The maturity level of a component may not be the same as in the previous one, as CSF for a component are not the same as for another, so the mastery of processes associated with each CSF must be assessed. We describe the proposed maturity model for the warehousing function in Moroccan companies in the table 2.

Table 2. Maturity Model for the warehousing function in Moroccan companies [39]

Component of the Maturity Model For Warehousing Function	CSFs for Warehousing Performance Improvement	Maturity level 1 (Initial)	Maturity level 2 (Defined)	Maturity level 3 (Managed)
Design of the warehouse	Definition of an optimal location of the Warehouse	The location of the Warehouse is not studied	The location of the Warehouse is defined	The location of the Warehouse is defined optimally
	Definition of an optimal design of the warehouse	The design of the warehouse is not studied	The design of the warehouse is defined	The design of the warehouse is defined optimally
	Definition of functional areas in the Warehouse	The functional areas are not defined in the Warehouse	The functional areas are defined in the warehouse	The functional areas are defined in the warehouse optimally
	Definition of warehouse storage capacity	The warehouse storage capacity is not defined	The warehouse storage capacity is defined	The storage capacity is considered in the design of the warehouse
	Definition of the developments of the warehouse storage capacity in the future	The evolution of the warehouse storage capacity in the future is not defined	The evolution of the warehouse storage capacity in the future is defined	The evolution of storage capacity in the future is considered in the design of the warehouse
Means for the warehouse	Selection of the appropriate storage systems	The storage systems are not used	The storage systems are used	Storage systems used are optimized
	Selection of the appropriate handling equipment	The handling equipment are not used	The handling equipment are used	Handling equipment used are optimized
	Definition of the necessary and efficient warehouse staff	The warehouse staff is not defined	The warehouse staff is defined	The warehouse staff is defined optimally

	Use of ICTs	ICTs are not used in the warehouse	ICTs are used in the warehouse	ICTs used in the warehouse are optimized
Operations in the warehouse	Definition of a rule for the allocation of products to the functional areas	The allocation of products to functional areas is done randomly	A rule for the allocation of products to the functional areas is defined	The allocation of products to the functional areas is optimized
	Definition of a rule for the allocation of products to the storage areas	The allocation of products to the storage areas is done randomly	A rule for the allocation of products to the storage areas is defined	The allocation of products to the storage areas is optimized
	Definition of a policy of picking	The picking policy is not defined	The picking policy is defined	The picking policy is optimized
	Definition of a policy of delivery	The delivery policy is not defined	The delivery policy is defined	The delivery policy is optimized
	Setting a policy for the return processing	The return processing policy is not set	The return processing policy is defined	The return processing policy is optimized
	Optimization of internal logistics for operations in the warehouse	Internal logistics for operations in the warehouse is done randomly	A rule for internal logistics for operations in the warehouse is defined	Internal logistics for operations in the warehouse is optimized
Management of the warehouse	Definition of inventory management policy	The inventory management policy is not set.	The inventory management policy is defined	The inventory management policy is optimized
	Definition of the staff allocation procedure to different missions	The staff allocation procedure for different missions is not defined	The staff allocation procedure to different missions is defined	The staff allocation procedure to different missions is optimized
	Definition of the quality control procedures in the warehouse	The quality control procedures in the warehouse are not defined	The quality control procedures in the warehouse are defined	The quality control procedures in the warehouse are used and improved continuously
	Definition of the security control procedures in the warehouse	The s security control procedures in the warehouse are not defined	The security control procedures in the warehouse are defined	The security control procedures in the warehouse are used and improved continuously
	Formalization of all working procedures	The working procedures in the warehouse are not defined	The working procedures in the warehouse are formalized	The formalized working procedures in the warehouse are used and improved continuously
	Use of a warehouse management information system	A WMS is not used in the warehouse	A WMS is used in the warehouse	The WMS used in the warehouse is optimized

4. CASE STUDY: MATURITY OF THE WAREHOUSING FUNCTION IN A MOROCCAN COMPANY

The proposed model has two objectives: first provide a framework to assess maturity level of warehousing function in Moroccan companies, and second, provide support to companies to develop a roadmap for improving their warehousing functions. To validate our model and examine whether it is suitable for those two uses, we achieve a case study in a Moroccan company. It was created based on an internship with a student engineer in 2015.

We choose SICOTREM Company because its top management needs an action plan to improve the warehousing function. Indeed, SICOTREM suffers from several logistical problems, namely: Delivery delays because of shortages; the work without added value repeated in its warehouse; losses from the current way of managing the warehousing function.

SICOTREM is the Moroccan leader in the production of spikes, wire annealing, wire mesh and galvanized wire. He has over 40 years of experience in this field and with a production capacity of about 2000 tons / month.

4.1 Implementation of the maturity model for the warehousing function

For the implementation of the maturity model, we propose an evaluation methodology based on questionnaires that will provide the maturity levels for each CSF, and therefore by aggregation for a component, and for the warehousing function. We propose defining a team composed of logistic manager and the responsible for each component to assess, perform the audit and recover as much as possible information for giving a reliable rating for the maturity that best reflects the current state of the warehousing.

The questionnaires are constructed according to the maturity levels defined for each CSF j (each CSF corresponds to three questions). These questions are generic and can be applied in all components. The difference in the assessment is made by the answers that will depend on the state of the CSF.

For each question, there are three possible answers: yes, no and does not apply. To get the maturity level of a CSF j for a component i, M CSF ij, we compare the responses to maturity scales (ranging from level 1 to 3). To obtain it, a positive answer to the question of the same level is necessary, as in most maturity models (CMMI 2006 [79]) it must have completely crossed a maturity level before proceeding to the next.

The results of the evaluation are used to identify areas for improvement in the components of the warehousing function through the comparison between the levels of maturity. When the warehousing function reaches a specified maturity level in a CSF, the improvement roadmap includes the next level. If level 3 is reached, the company must keep it.

4.2 Maturity of the SICOTREM's Warehousing function

The maturity level of the SICOTREM's warehousing function is summarized in table 3 as it was assessed by SICOTREM's logistics manager.

Table 3. The maturity level of SICOTREM's Warehousing Function

Component of the Maturity Model For Warehousing Function	CSFs for Warehousing Performance Improvement	Maturity Level of SICOTREM's CSFs for Warehousing Performance Improvement	Maturity Level of SICOTREM's Component of the Maturity Model For Warehousing Function
Design of the warehouse	Definition of an optimal location of the Warehouse	Level 1	Level 1
	Definition of an optimal design of the warehouse	Level 1	
	Definition of functional areas in the Warehouse	Level 1	
	Definition of warehouse storage capacity	Level 1	
	Definition of the evolution of the warehouse storage capacity in the future	Level 1	
Means for the warehouse	Selection of the appropriate storage systems	Level 1	Level 1
	Selection of the appropriate handling equipment	Level 2	
	Definition of the necessary and efficient warehouse staff	Level 2	
	Use of ICTs	Level 1	
Operations in the warehouse	Definition of a rule for the allocation of products to the functional areas	Level 1	Level 1
	Definition of a rule for the allocation of products to the storage areas	Level 1	
	Definition of a policy of picking	Level 1	
	Definition of a policy of delivery	Level 1	
	Setting a policy for the return processing	Level 1	
	Optimization of internal logistics for operations in the warehouse	Level 1	
Management of the warehouse	Definition of inventory management policy	Level 1	Level 1
	Definition of the staff allocation procedure to different missions	Level 1	
	Definition of the quality control procedures in the warehouse	Level 1	
	Definition of the security control procedures in the warehouse	Level 1	
	Formalization of all working procedures	Level 1	
	Use of a warehouse management information system	Level 1	
Maturity level of the SICOTREM Warehousing function			**Level 1**

4.3 Roadmap for improving SICOTREM's warehousing performance

After the assessment of maturity level, based on the proposed model, we developed an improvement roadmap for SICOTREM Company's warehousing function as it is shown in table 4.

Table 4. The improvement roadmap for SICOTREM's Warehousing function

Component of the Maturity Model For Warehousing Function	CSFs for Warehousing Performance Improvement	Improvement roadmap for SICOTREM's Warehousing function
Design of the warehouse	Definition of an optimal location of the Warehouse	The location of the Warehouse must be defined
	Definition of an optimal design of the warehouse	The design of the warehouse must be defined
	Definition of functional areas in the Warehouse	The functional areas must be defined in the warehouse
	Definition of warehouse storage capacity	The warehouse storage capacity must be defined
	Definition of the evolution of the warehouse storage capacity in the future	The evolution of the warehouse storage capacity in the future must be defined
Means for the warehouse	Selection of the appropriate storage systems	Definition of the storage systems to be used in the warehouse
	Selection of the appropriate handling equipment	Optimization of the handling equipment used
	Definition of the necessary and efficient warehouse staff	Optimization of the warehouse staff
	Use of ICTs	Definition of the ICTs to be used in the warehouse
Operations in the warehouse	Definition of a rule for the allocation of products to the functional areas	A rule for the allocation of products to the functional areas must be defined
	Definition of a rule for the allocation of products to the storage areas	A rule for the allocation of products to the storage areas must be defined
	Definition of a policy of picking	The picking policy must be defined
	Definition of a policy of delivery	The delivery policy must be defined
	Setting a policy for the return processing	The return processing policy must be defined
	Optimization of internal logistics for operations in the warehouse	A rule for internal logistics for operations in the warehouse must be defined
Management of the warehouse	Definition of inventory management policy	The inventory management policy must be defined
	Definition of the staff allocation procedure to different missions	The staff allocation procedure to different missions must be defined
	Definition of the quality control procedures in the warehouse	The quality control procedures in the warehouse must be defined
	Definition of the security control procedures in the warehouse	The security control procedures in the warehouse must be defined
	Formalization of all working procedures	The working procedures in the warehouse must be formalized
	Use of a warehouse management information system	A WMS must be used in the warehouse

5. Conclusion

Based on the critical success factors and the concept of maturity, a four-component Maturity Model for the warehousing function in Moroccan companies has been developed. The suggested Maturity Model makes it possible for companies to identify, explain, assess and improve this critical function in a supply chain.

The results obtained from the case study to validate the proposed model are interesting. They allowed assessing the maturity level of the warehousing function and developing a roadmap for improving its performance. Indeed, the diagnostic results confirmed the concern of top management of the company subject of the case study likened to failures of the warehousing function. The "initial level" obtained for the case of this company shows that more actions need to be implemented to achieve higher levels of maturity. The roadmap obtained trace the priority improvements that need to be applied to achieve at least level two of maturity.

For future work, the maturity model suggested should be discussed with other empirical studies in different contexts.

References

[1] Baker, P. (2007). An exploratory framework of the role of inventory and warehousing in international supply chains. The International Journal of Logistics Management, 18(1), 64-80.

[2] Ramaa, A., Subramanya, K. N., & Rangaswamy, T. M. (2012). Impact of warehouse management system in a supply chain. International Journal of Computer Applications, 54(1).

[3] Min, H. (2006). The applications of warehouse management systems: an exploratory study. International Journal of Logistics: Research and Applications, 9(2), 111-126.

[4] Ibbs, C. W., Reginato, J. M., & Kwak, Y. H. (2004). Developing project management capability: Benchmarking, maturity, modeling, gap analyses, and ROI studies. The Wiley guide to managing projects, 1214-1233.

[5] Cooke - Davies, T. (2004). Project management maturity models. The Wiley guide to managing projects, 1234-1255.

[6] Qureshi, T. M., Warraich, A. S., & Hijazi, S. T. (2009). Significance of project management performance assessment (PMPA) model. International Journal of Project Management, 27(4), 378-388.

[7] Jens Pöppelbuk, Maximilian Röglinger, "What makes a useful Maturity Model? A Framework of general Design Principals for Maturity Models and Its Demonstration In Business Process Management", ECIS 2011 PROCEEDINGS.

[8] Bititci, U. S., Garengo, P., Ates, A., & Nudurupati, S. S. (2015). Value of maturity models in performance measurement. International journal of production research, 53(10), 3062-3085.

[9] K. McCormack, MB. Ladeira, and MPV. Oliveira, "Supply Chain Maturity and Performance in Brazil", Supply Chain Management, Vol. 13, No 4, 2008, pp. 272-282.

[10] H.G. Reyes, and R. Giachetti, "Using experts to develop a supply chain maturity model in Mexico", Supply Chain Management, Vol. 15, No. 6, 2010, pp. 415-424.

[11] T. Wettstein, and P. Kueng, "A maturity model for performance measurement systems", in: C.A. Brebbia, P. Pascolo, (Eds.), Management Information Systems GIS and Remote Sensing, WIT Press, Southampton, 2002.

[12] Hong, Y., & Zhong-Hua, Y. (2013). Supply chain dynamic performance measurement based on BSC and SVM. IJCSI International Journal of Computer Science Issues, 10(1), 271-277.

[13] Labiad, N., Beidouri, Z., & Bouksour, O. (2012). Evaluation of coordination contracts for a two stage supply chain under prices dependent demand. IJCSI International Journal of Computer Science Issues, 9(1), 151-159.

[14] Longo, F. (2011). Operational Strategies and Internal Logistic Costs Analysis in a Real Warehouse Based on Modeling & Simulation. IJCSI International Journal of Computer Science Issues, Vol. 8, Issue 4, No 1, 39-47.

[15] Baker, P. (2007). An exploratory framework of the role of inventory and warehousing in international supply chains. The International Journal of Logistics Management, 18(1), 64-80.

[16] European Logistics Association / AT Kearney (2004), Differentiation for Performance, Deutscher Verkehrs-Verlag GmbH, Hamburg.

[17] Establish Inc / Herbert W. Davis & Co. (2005), Logistics Cost and Service 2005. Council of Supply Chain Management Professionals Conference presentation. Accessed at www.establishinc.com on 23rd August 2006.

[18] Frazelle, E.H. (2002), Supply Chain Strategy: the Logistics of Supply Chain Management, McGraw-Hill, New York.

[19] Werling, Chris. (2006). Tomorrows Warehouse, Corner Stone Solutions, Inc.

[20] Öztürkoğlu, Ö., Gue, K. R., & Meller, R. D. (2014). A constructive aisle design model for unit-load warehouses with multiple pickup and deposit points. European Journal of Operational Research, 236(1), 382-394.

[21] Johnson, A., & McGinnis, L. (2010). Performance measurement in the warehousing industry. IIE Transactions, 43(3), 220-230.

[22] Les Guides CGEM: Guide de la qualité pour les PME-PMI. (December 2009).

[23] Naciri, A. B. (2010). La contribution de la logistique à l'intégration des sous-traitants marocains du secteur textile-habillement à la logique des donneurs d'ordre étrangers. Repenser les économies africaines pour le développement, 121.

[24] La Logistique du Commerce et la Compétitivité du Maroc. Banque Mondiale & Ministère de l'Equipement et du Transport. (Mai, 2006).

[25] Le secteur logistique sur la rive sud de la Méditerranée Occidentale. CETMO. (2010).

[26] Balambo, M. A., & Houssaini, A. (2011). Typologies d'intégration des Supply Chains: perspectives pour le contexte Marocain. In LES DEFIS DU MANAGEMENT INTERNATIONAL A L'AUBE DU XXIème SIECLE (p. 21).

[27] Stratégie Nationale de développement de la compétitive logistique. Ministère de l'équipement, du transport et de la logistique, Maroc. (2010).

[28] Mohan, V. E. (2012). Warehousing and Inventory Management. CII Institute of Logistics. Chennai: www. ciilogistics.com.

[29] Razik, M., Radi, B. & Okar, C. (2016). Les facteurs clés de succès pour l'amélioration de la performance de la fonction entreposage au sein des entreprises marocaines. Ouvrage collectif "Management logistique au Maroc".

[30] Basque R. – 2004 – « CMMI : Un itinéraire fléché vers le Capability Maturity Model Intregration » - Dunod

[31] Patnayakuni, R., Rai, A. & Tiwana A. – 2007 – « Systems Development Process Improvement: A Knowledge Integration Perspective »

[32] Simpson, J.A. and E.S.C. Weiner, The Oxford English Dictionary. 2 ed. 1989, Oxford, UK: Oxford University Press.

[33] Fraser, P., J. Moultrie, and M. Gregory, The Use of Maturity Models/Grids as a Tool in Assessing Product Development Capability. in IEEE IEMC 2002. 2002. Cambridge, UK.

[34] Chafik Okar, Zitouni Beidouri, Said Mssassi, Said Barrijal, "A maturity model for SCPMS project: an empirical investigation in large sized Moroccan companies", IJCSI International Journal of Computer Science Issues, Vol. 8, Issue 2, March 2011 ISSN (Online): 1694-0814 www.IJCSI.org.

[35] T. J. Cooke-davies, and A. Arzymanowc, "The maturity of project management in different industries an investigation into variations between project management models", International Journal of Project Management, Vol. 21, 2003, pp. 471-478.
[36] S. B. Abd Karim1, R. Adawiyah, A. Rahmi, M. Suhaimi Mohd Danuri3, O. Mohamed, " Developing the Value Management Maturity Model (VM3©)", Journal of Design and Built Environment Vol. 14(1), June 2014.
[37] A. F. Bay, and M. Skitmore, "Project management maturity: some results from Indonesia", Journal of Building and Construction Management, Vol. 10, 2006, pp. 1-5].
[38] Irena Hribar Rajterič, "Overview of Business Intelligence Maturity Models", Management, Vol. 15, 2010, 1, pp. 47-67
[39] Razik, M., Radi, B. & Okar, C. (2016). Vers un model de maturité de la fonction entreposage chez les entreprises marocaines. Conférence Journée doctorants FST de Settat. Mai 2016.

DOI: 10.1051/matecconf/201710500008

A Middleware Architecture for RFID-enabled traceability of air baggage

T. Bouhouche[1], A. Raghib[2], B. Abou El Majd[2], M. Bouya[3], M. Boulmalf[1]

[1] *ITC Lab., International University of Rabat, Morocco*
[2] *LIMSAD Lab., Hassan II university of Casablanca, Morocco*
[3] *REMA Lab., International University of Rabat, Morocco*

Abstract. 1980 marked the start of a boom in radiofrequency identification (RFID) technology, initially associated with a growing need for traceability. In view of the technological progress and lower costs, RFID's area of application became much broader and, today, multiple business sectors take advantage of this technology. However, in order to achieve the maximum benefits of RFID technology, the data collected should be delivered in the best conditions to the whole applications that have need of its exploitation. For that, a dedicated middleware solution is required to ensure the collection of RFID data and their integration in information systems. The issues and key points of this integration as the description of the RFID technology will be summarized in the present paper, with a new middleware architecture. We focus mainly on components and the design of our middleware MedRFID, solution developed in our Lab, and which integrates mobility and provides extensibility, scalability, abstraction, ease of deployment and compatibility with IATA standards and EPCglobal standards. Moreover, we have developed an application (FindLuggage) allowing a real time tracking of luggage in the airport, based on the proposed middleware MedRFID.

1 Introduction

RADIO FREQUENCY IDENTIFICATION, or RFID, is a generic term for technologies that use radio waves to automatically identify people or objects [16]. This technology has recently seen growing interest from a wide range of industries such as retail, pharmaceutical, and logistics [18-21]. In these domains, RFID technology holds the promise eliminating many existing business problems by bridging the economically costly gap between the virtual world of IT systems and the real world of products and logistical units. Common benefits include more efficient material handling processes, elimination of manual inventory counts, and the automatic detection of empty shelves and expired products in retail stores. RFID technology has a number of advantages over other identification technologies. It does not require line of-sight alignment, multiple tags can be identified almost simultaneously, and the tags do not destroy the integrity or aesthetics of the original object. The location of tagged objects can thus be monitored automatically and continuously.

In traditional RFID applications, such as access control, there was a little need for an RFID middleware because the RFID readers were not networked and the RFID data were only consumed by a single application. The appearance of the first RFID middleware on the market enables the integration of RFID technology in various fields (Air, Logistic, etc). Nevertheless, the fact remains that they allow the identification of several technical issues regarding this integration. In novel application domains, such as supply chain management and logistics, there is no longer a 1-to-1 relationship between reader and application instance. However, in these domains, many readers distributed across factories, warehouses, and distribution centers capture RFID data that need to be disseminated to a variety of applications. This introduces the need for an RFID infrastructure that hides proprietary reader device interfaces, provides configuration and system management of reader devices, and filters and aggregates the captured RFID data.

This paper has been organized in the following way: The first part is comprised of a presentation of the RFID systems. A description of the proposed MedRFID and the FindLuggage application are presented in the second part. The second and final part concludes the paper.

2 RFID Systems

The RFID systems basically consist of two or three elements: a tag/transponder and a reader for a Simplified RFID system, or a tag/transponder a reader and a middleware deployed at a host computer. The RFID tag is a data carrier part of the RFID system, which is placed on the objects to be uniquely identified. The RFID reader is a device that transmits and receives data through radio waves using the connected antennas. Its functions include powering the tag, and reading/writing data to the tag. As shown in Fig. 1, the signals sent by the reader's antennas form an interrogation zone made up of an electromagnetic field. When a tag enters this zone, it gets activated to exchange data with the reader [2]. Later, the identification data read by the RFID reader is processed by the software system, known as the RFID middleware. The RFID middleware manages readers, as well as filters and formats the RFID raw tag data so that they can be accessed by the various interested enterprise applications [11]. Hence, the middleware is a key component for managing the flow of information between tag readers and enterprise applications [3].

Major advantages of using RFID as an auto-ID system are the following:

• RFID readers do not require a line of sight to access data from the RFID tags.

• RFID systems can read data over varied range from few centimeters to few hundred meters.

• RFID readers can interrogate, and make RFID tags readings much faster.

• RFID systems can read and write different sizes of data from reader / to the tag, based on the type of tag.

• RFID systems can read tags in harsh environments, without any human interference.

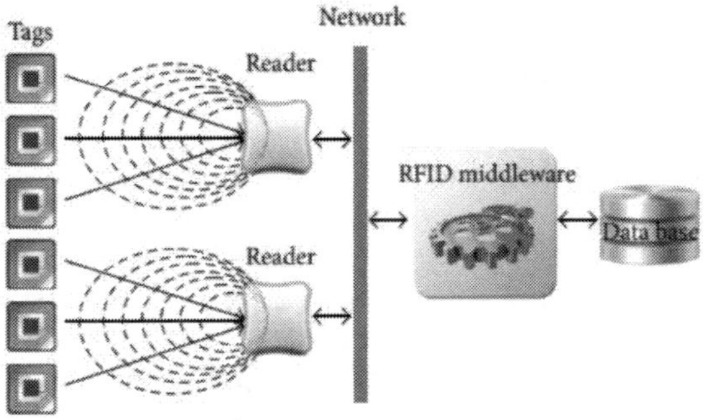

Figure 1. RFID System components.

Indeed, since the introduction of RFID technology in global solutions, a recurring issue is the integration of RFID in a transparent and efficient manner. As a result, there is a strong mobilization on the issue of research and development in mainly three areas:

➢ *Objects Identification:*

For a better identification of object, we need to focus the development on a standardization of RFID tags and a good transcription codes on RIFD tags (Algorithm).

➢ *Communication between objects and middleware:*

The start point for this part is a good choice of useful RFID technologies, after a development of communication protocols between readers is necessary, following the middleware standardization can help us to done a good follow up with the objects, then the normalization of protocols shall needed for the better lopping of the communication steps.

➢ *The information processing:*

Decoding the RFID information collected (standard, algorithm) is important for structuring collected RFID information to make it usable directly by any application of the information system, which need a real Time treatment volume (architecture allows scaling mass treatment of RFID data collected).

A. RFID System Components:

RFID systems are produced by many manufacturers and exist in countless variants. However, a RFID system consists mainly of three components; the transponder/tag, reader, and RFID middleware.

1) RFID Transponder/Tag:

A RFID transponder, or tag, consists of a chip and an antenna [22]. A chip can store a unique serial number or other information based on the tag's type of memory. The tag's type of memory can be read-only, read-write, or write-once and read-many [11]. Read-only tags are much cheaper to produce and are used in most current applications. Read-write tags are useful when information needs to be updated [2]. The antenna is used to transmit information from the chip to the reader, and the larger the antenna the longer the read range. The RFID tag can be either attached or embedded in an object to be identified, and can be scanned by mobile or stationary using radio waves [11, 14-17].

2) RFID Reader:

A RFID Reader is a scanning device that reliably reads the tags and communicates the results to the middleware. A reader uses its own antennae to communicate with the tag by broadcasting radio waves to which all tags within range will respond. Readers can process multiple items at once, allowing for increased read processing times. They can be either mobile or stationary, so that their optimal employment in the area can be found [12-13], and they are differentiated by their storage capacity, processing capability, and the frequency they can read [11].

Different designs of readers exist, because different applications have different requirements. RFID readers are classified into three types [2]:

• OEM readers: Original Equipment Manufacturers readers are mostly used for data capture systems, access control systems, and robots.

• Industrial use readers: used in assembly and manufacturing plant.

• Portable readers: These readers are more mobile than the other readers, and supported with a LCD display and keypad. This kind of readers is used in animal identification, device control and asset management applications.

3) RFID Middleware :

The middleware refers broadly to software or devices that connect RFID readers and the data they collect to enterprise information systems. RFID middleware helps make sense of RFID tag reads, applies filtering, formatting and logic to tag data captured by a reader, and provides this processed data to back-end applications [3]. RFID middleware serves in managing the flow of data between tag readers and enterprise applications, and is responsible for the quality, and therefore usability of the information. It provides readers with connectivity, context-based filtering and routing, and enterprise/B2B integration. RFID middleware design and components will be discussed further in the next sections.

When designing a RFID middleware solution, the following issues need to be considered:

- **Multiple hardware support:** The middleware must provide a common interface to access different kinds of hardware offering different features.
- **Synchronization and scheduling:** There should be intelligent scheduling and synchronization among all the processes of the middleware. This minimizes the latency and improves the efficiency of the middleware.
- **Real-time handling of incoming data from the RFID readers**: The middleware should handle the huge amount of data captured by the connected readers in real time without read misses.
- **Interfacing with multiple applications:** The middleware should be capable of interacting with multiple applications simultaneously, by catering to all the requirements of the applications with minimal latency.
- **Device neutral interface to the applications:** The application developer should only use the generic set of interfaces provided by the middleware independently of the type of hardware connected to the system.
- **Scalability:** The middleware design must allow easy integration of new hardware and data processing features.

B. RFID Middleware Components:

A RFID middleware is the interface that sits between the RFID hardware and RFID applications. It provides the following advantages:

- It hides the RFID hardware details from the applications.
- It handles and processes the raw RFID data before passing it as aggregated events to the applications.
- It provides an application level interface for managing RFID readers and querying the RFID data.

A layer of the RFID middleware incorporates all the device drivers of different hardware and exposes the application standard interfaces to access this hardware. If the application was provided with all the device drivers of all connected readers, it will be a hard job to manage and interface each of the devices. The application developer will then need to understand all the hardware-specific internals and operations. Also, the application, if provided with the huge amount of raw tag data reported by the readers, will find it very difficult to process the data in real time. A RFID middleware provides a standardized way of dealing with this flood of information, which processes the raw data and provides the application with clean and filtered data.

RFID middleware is generally composed of four major layers:

➢ *Reader Interface:*

The reader interface is the lowest layer of the RFID middleware which handles the interaction with the RFID hardware. It maintains the device drivers of all the devices supported by the system, and manages all the hardware related parameters like reader protocol, air interface, and host-side communication.

➢ *Data Process or and Storage:*

The data processor and storage layer is responsible for processing and storing the raw data coming from the readers. Examples of processing logic carried by this layer are data filtering, aggregation, and transformation. This layer also processes the data level events associated with a specific application.

➢ *Application Interface:*

The application interface provides the application with an API to access, communicate, and configure the RFID middleware. It integrates the enterprise applications with the RFID middleware by translating the applications" requests to low level middleware commands.

➢ *Middleware Management:*

The middleware management layer helps managing the configuration of the RFID middleware, and provides the following capabilities:

- Add, configure, and modify connected RFID readers.
- Modify application level parameters such as filters, and duplicate removal-timing window.
- Add and remove services supported by the RFID middleware.

RFID readers are typically abstracted as a logical reader which is either a collection of several readers or a part of the reader. This grouping mechanism is used where there is a need to have a set of readers capturing data from a particular area such as a warehouse with many loading docks. The advantage of this is that the application can query a small number of logical readers rather than having to aggregate events from each of the individual readers.

There are two standardized interaction models used to define the communication between the middleware and the applications. An application can operate at synchronous mode when requesting services on demand or asynchronous mode when it registers for information to be sent to it when certain conditions are met. RFID middleware usually provide some kind of data filtering, because sometimes it might be required to report only certain type and value of the tag data to the application. The application must provide a set of defined patterns to the middleware. The middleware then allows only data that matches the pattern to be reported to the application. For example, if an application must only see tag data that starts with a specific pattern such as "XYZ20", the filter can be set to this value by the application and communicated to the middleware [2].

3 MEDRFID

3.1 Introduction: Why MedRFID?

Because some existing RFID middleware use a standard functionalities, our concern is to design and implement a specific middleware platform called "MedRFID, which is distinguished by its innovative aspects in terms of architecture including mobility to enable the collection of information from mobile RFID readers, operating aspects of the data in real time with the ability to generate a business rules engine and especially by integrating communication protocol RFID readers disregarding the dependency to the initial manufacturer protocol. The implementation of this innovative architecture will give real flexibility for users of the middleware and above all a means of real-time processing a large volume of data that emanate from heterogeneous environments while ensuring ease of integration of these data and information and their best use by business applications systems.

3.2 MedRFID: Design and Architecture

RFID middleware is the core in the implementation of RFID technology in companies that have large amount of data to be processed and, above all, to share with the various components of the value chain and with its partners.
Thus, the successful integration of RFID within companies is conditioned by the successful design and implementation of RFID middleware. That is why our research team chose to design and implement a RFID middleware which integrates the latest innovations in the field.
MedRFID is a middleware that has been designed to meet the overall functionality that must contain a conventional RFID middleware, but it also contains areas of innovation mainly around the mobility and independence of manufacturers. It is an innovative device management and event processing platform at the edge of the enterprise. It is designed to provide a scalable, extensible platform for deployment, and management of rich RFID and sensor solutions.

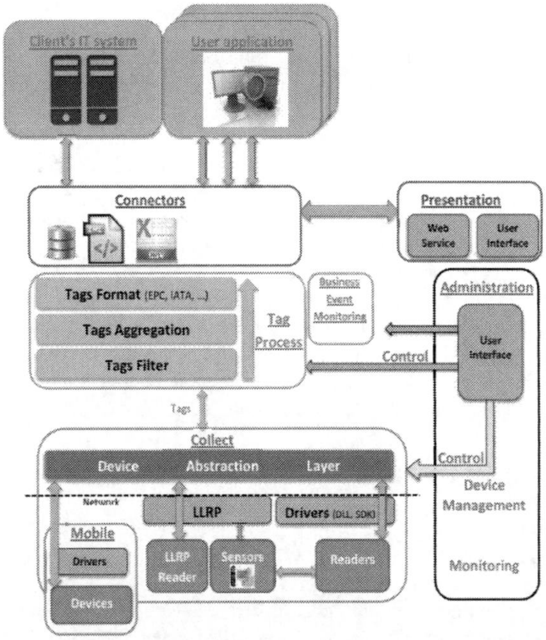

Figure 2. MedRFID Architecture

As shown in Fig. 2, the MedRFID architecture is composed of seven major layers

2.1) Collect:

The collect module is the heart of the middleware. It's responsible of all collected tag's data, and it's a key instrument of the connection with the readers and the software intelligence. Moreover, it's used to control the readers, add data to the tags and format them in a single format regardless the used reader. To control the readers, this module has embedded an LLRP driver but also each driver needed for RFID readers, which don't support LLRP. These drivers are often provided by RFID manufacturer in different programming languages.
Each tag is formatted in an abstract layer in order to have the same format for all the tags whether it come from a mobile device, a LLRP or a specific reader (manufacturer driver).

2.1.1) LLRP:

LLRP (Low Level Reader Protocol) is a standard that facilitate the communication between the software and the readers. All the RFID readers speak the same language, so they can all understand the software whichever their types. This allows to the customers the choice of its hardware with the software. If they want to change the hardware, it could be without affecting the software solution. The collect process is as follow:
The software sends to the readers a RO Spec (Reader Operation spec). The RFID readers use these specs when they are enabled and when the start condition is reached. Then, the reader sends to the software a report of each tag or a list of them.

2.1.2) Custom Drivers:

Some readers do not support LLRP, but these readers may be interesting to support. Indeed, some of them are really good in RFID performance. Fortunately, readers manufacturer provide most often the driver in different languages to communicate with the reader. Languages mostly supported are C, C# and JAVA. So, to implement the driver for these readers, we have to develop a separate module which is an interface between the driver provided by manufacturer and the command provided by the middleware.

2.1.3) Mobile Readers:

Some of the mobile readers fully support LLRP and they can be used like fixed readers. For others, there are also drivers provided by manufacturers for control like the fixed RFID readers. These drivers can be embedded in a mobile app and transmit data to the middleware backend when it is connected to the cellular network.

2.1.4) Sensors:

The middleware should also support Sensors. There are two types of sensors: IN Sensors and OUT Sensors.
IN Sensors are detectors, radars and are used to trigger the collection of readers. For example, when a tracked object is coming, IN Sensors detect it and trigger the start of the collection. With these sensors, it is possible to read only when there is something to track and facilitate the process (hardware + software).
OUT Sensors are used by users on the ground. It can be for example lights or alarms. They are used to secure the RFID process, for example, a green light may be switched on when a tag is read to show to the user that it can continue. Or a red light can be switched on if there is a problem with the tag read or if an access is forbidden to this user.

2.2) Mobile Application:

The software should include a mobile application. Mobile RFID readers allow RFID collect stationary objects. The reader is embedded in a mobile device with powerful RFID antennas and can communicate with the backend over cellular network.
Tags are stored directly in database if a connection is established with the backend or the mobile application stock tags read in the local memory if not. Like the collect module, tags are formatted in the same way in an abstract layer.
For the backend, reader device is transparent whether it is from a fixed reader or a mobile device. A mobile application is useful for huge objects to track and it is not possible to read them with a fixed reader.

2.3) Administration:

The administration module concerns the settings of the middleware. RFID Readers, antennas and sensors settings can be modified such as IP Address, Power, and numbers of antennas… In addition, each reader's RO Specs can be altered: Start/Stop trigger, numbers of tag reports…
Tag process preferences can also be modified, for example: Duration of the duplicate elimination process, decoding process…. Finally, Business rules can be added, edited or removed. Reader supervision will also be embedded in this module.
The supervision is a module which consists of logging each event about a reader. When something does not work properly, the middleware should send an alert (via mail, messages, SMS …) to explain where the problem is.
Problems can be about the network, the reader, or its antennas. Ideally, the middleware should explain what type of alert it is and what the problem is. This is possible by doing multiple tests: For example, if there is no tag reads for X hours but the ping to the reader is OK, we may think it's a problem in the reader or its antennas (or both) but not a network issue.

2.4) Tag Process:

Each tag will go through a tag process that can have multiple operations such as filtering, eliminating duplicates, time stamping, contextualization and/or formatting.
The tag process is composed of three layers:

- *Tags Filter:*

The Tags filter is the layer which removes tags that we don't want to read. Each tag has a Unique ID and a User Memory where you can write personal data. This memory is often used to filter tags because every tag that we are interested in contains this value in its User Memory.

- *Tags Aggregation:*

The Tags aggregation is the mapping between the Tag Number (UID) and the business data we have in a database. The tag number is just an ID but it's often (always) linked to personal or business data. For example, the box with the RFID N°XXX (UID) contains three of YYY items (Business Data).

- *Tags Format:*

In some case, some tags are encoding following a strict format. So, the number contains data. There are multiple formats for RFID Tags. The middleware should include encoding and decoding algorithm in order to get back the information.

2.4.1) Duplicate Elimination:

The duplicate elimination is very important in an RFID application. It's often the first layer in the tag process in a middleware.
RFID readers can read more than 100 tags per second. Indeed, the same tag is read more than once but only one read is important during this second. So the duplicate elimination process permits to keep only one read of a tag during a customizable duration. With the duplicate elimination filter, the middleware eases the database from storing duplicate data. So, instead of 300 reads/writes in 3 seconds, there are only 3. Each tag will then be evaluated in a filter layer and then a business rules engine to redirect it if needed.

2.4.2) Filtering:

A filter defines rules that allows or stops tags during the tag process. Sometimes, multiple actors in the same location use RFID tags. In order to only keep the interesting tags, filter rules allows removing tags that we don't need to read. Filtering can be on specific serial numbers of tags, or in business number of a tag. Some tags are encoding following different norms. These norms such as IATA RP1740c or GRAI-96 encode business data in the tag. For example, IATA RP1740c encodes data about flights, so in the tag number is encoded: the company prefix, the date time of the flight, source and destination airports etc… By decoding the tag, it is possible to filter tags that have the company prefix for AIR France or other companies.
If a tag match the filter value, then the tag will be saved and continue through the tag process. Else, it is deleted

from the tag process. With the filtering, only interesting tags are collected and saved in the database.

2.4.3) Business Rules Engine:

Business rules are rules that match some tags or dates and practice different services according the tag. This allows for example to send data to a client for all the tags of its client but hide him the rest of the reads. It is possible to edit and add new rules to the engine. Then, rules are affected to the tags read events through the engine.

2.5) Connectors:

Tag reads, once being treated in the tag process and by the business rule engine, will then be transmitted via connectors such as Database connector, Files or Queues. The connectors allow external application to get the data provided by the middleware. External applications may be presentation software in order to show statistics, tag reads, etc. It can also be formatted files such as comma-separated files or web services in order to send the data to another software in order to automatically fulfill stocks etc.

2.6) The User Interface:

This is the view of the middleware. It is composed of three tabs: Reads, Writes and Administration.

The UI of the middleware shows tag reads in real time and permits the configuration of RFID readers / application.

The "Reads" view provides information about reads, EPC, date time, reader, antennas and ensures that the tag has been written into the database.

The user interface can also be developed by using the web service API.

The "Administration" view represents all the settings alterable for the middleware.

Reader settings: It comprises all the settings for each reader. How a read is triggered, how many antennas are connected to the reader, how long is the duplicate elimination.

The middleware settings represent all the settings alterable for the middleware itself: Connect or disconnect readers to the middleware, the SQL connection settings, and paths to different log files.

2.7) Client's IT System:

Each client of the middleware can be connected to it with the connectors and add data to its information system.

The middleware provides different connectors to transmit data to other information system. So, it is possible for external clients to develop a little interface in order to add the RFID data to their existing application or develop a new application by using these connectors.

The client's IT system is not a part of the middleware but we have to think it may be connected to it.

4 FIND LUGGAGE APPLICATION

With tens of millions of passengers traveling through the major European cities each year, through hundreds of check-in desks, luggage is extremely important for these airports. Traditional moving-belt bar code scanners require 'line of sight' in order to correctly read labels, but these could be hidden as the luggage moves on conveyor belts, rendering the luggage unidentifiable. This could slow the process down, as manual intervention is required to obtain the information and locate the luggage, and could directly contribute to misplaced or lost items.
In the airline industry, RFID considered as one of the best technologies that help the sector to remain competitive and innovative and to face the demands of increasingly cost and security, and increased traffic. This, to accommodate the needs of travelers but also to be operational.

Travelers are always afraid of luggage loss or theft. Now, RFID helps airports and airlines to limit this kind of risk, and meet the growing demands in the field of luggage handling (speed, security, cost control, etc.). As for industries, they benefit from changes in the direction of lower costs, faster processes, and improved communication.

The actors that interact directly with the system studied are:

- Holders of RFID tags: The luggage.
- RFID Readers: responsible for receive data, activation of RFID tags in each area, and communicate data to the middleware (MedRFID).
- MedRFID: collecting and sorting the data transmitted by the readers, to share with relevant the application or store them in a database.
- Employee: who may be responsible or single control officers who watch over the path of luggage to aircraft in good conditions, otherwise make alerts or even answer questions dedicated to specific luggage.

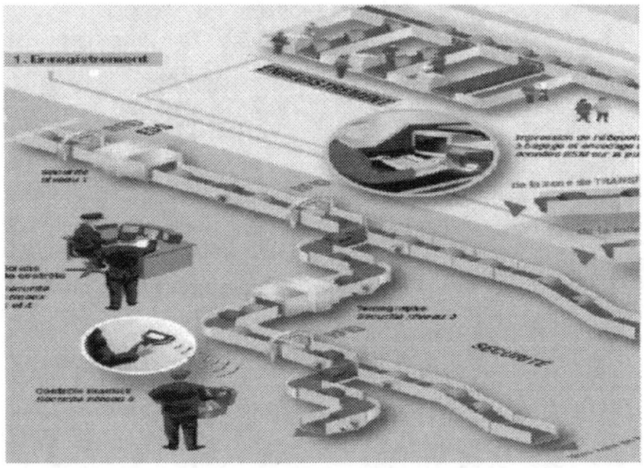

Figure 3. Example of platform tracking luggage in the airport.

Our study consists of the management and reporting of the processing of luggage immediately following registration (check-in). The RFID tag is already encoded and contains luggage's data, printed and stored on it. Then, by using MedRFID, we will track the luggage at any time, and any different points in passages (the inputs / outputs of luggage sorting areas, security & control, loading / unloading, etc.) as shown for example in figure 3, which allows an automated traceability, reliable and in real-time for a good geo-location.

To properly manage followed, we offer an application that will be the RFID interface solution installed and designed for both the environment and luggage, in order to track, locate luggage in real time, and ensure safe movement of flights, as well as avoiding wastage of time lost in inventories.

When a passenger makes the check-in step, an RFID tag is attached on his luggage containing specific information (ticket number, destination, etc.). Exploitation of this tag allows identifying his luggage at various traffic areas along its presence in the airport.

After the passage of luggage in an area, the RFID reader passes to the tag an enough energy to its activation, with the aim to read its information or to add new data, by interacting in both cases with the MedRFID.

After receiving the luggage data, MedRFID is responsible for centralizing, sorting and filtering it for the needs of business applications, by providing understandable and reliable data to the other applications or database. Therefore, the user can easily locates a specific luggage, checks the safety, sorts, and even monitors it in a very small time. Our application deals with actual and real requirements by identifying the objectives, priorities, rules of management and key processes of the organization as shown in figure 4.

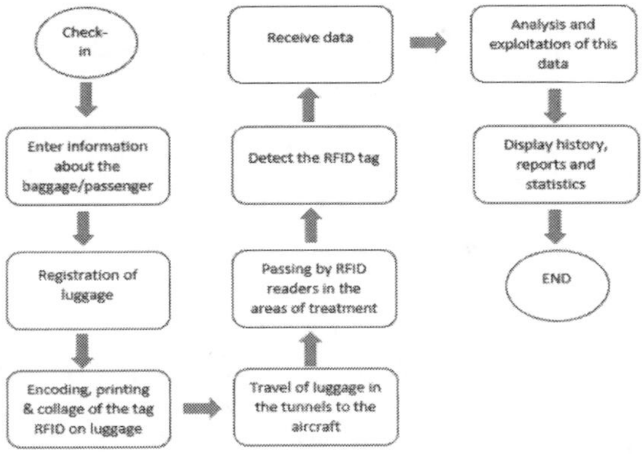

Figure 4. Luggage functional process.

The use of RFID traceability optimizes all stages of the processing airport, from check-in to loading bunkers namely: Registration luggage, security checks, sorting, reconciliation and loading. To overcome these problems and automate tracking of luggage, we decided to design and implement a web application: FindLuggage, based on the RFID solution and using the middleware MedRFID for tracking and monitoring a luggage in real time, and reducing the costs.

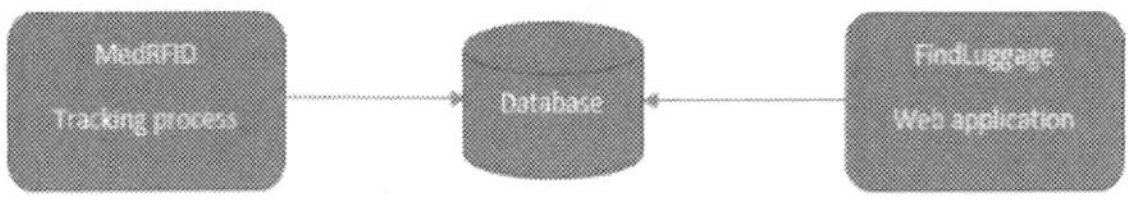

Figure 5. Communication between MedRFID and FindLuggage.

As shown in figure 5, the use of this service requires the implementation of a RFID reader at different points of traceability, and the placement of each piece on an RFID tag; this is followed by the use of MedRFID luggage tracking to collect information on these different points of traceability at the airport and save them in a database. Latterly, the application plays a role in integrating these data and providing information about the luggage. RFID readers can be deployed at both the check-in and the luggage collection stages, which enable to read the attached tags as it left the airport or, prior to collection by passengers. After the deployment of RFID readers, the middleware MedRFID collects traceability information on luggage at all tracking points within the airport and makes them available to the application via a database. The application FindLuggage operates and integrates these data and then uses them to be able to indicate the pathway of each luggage through the airport. It allows to find a misplaced luggage or give more details on the passage of a bag at a given collection point.

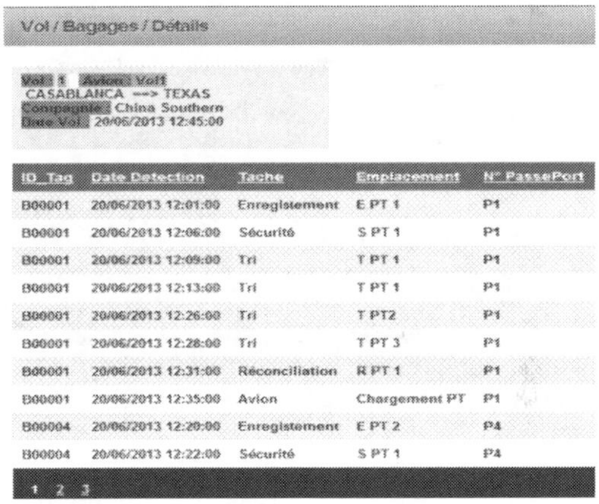

Figure 6. MedRFID application : FindLuggage.

Figure 7. MedRFID application : FindLuggage.

As show in figures 6 and 7, the FindLuggage application facilitates access to information for the employee according to its type (lead or manager). He has the possibility to track the complete path of all flight's luggage or a specific one, using the flight number or the identifier of a / set of RFID tag. This possibility is also allowed to the passengers in order to have a vision on their luggage, but with very limited access to the consultation option. MedRFID's FindLuggage application has a several attractive features:

- It improves traceability due to better-read rates.
- An easy integration of many readers, allowing a real time tracking of luggage.
- Reducing labor costs.
- Reducing the number of lost or delayed luggage.
- Reducing the luggage handling time.
- Larger encoding capabilities: Passenger/ Luggage data traveling with luggage.
- Improves the comfort and safety of people in transit, but also to simplify their movement within airports.
- To have a global vision on luggage present in a specific area and manage it in real time.
- Generate reports and statistics of usage history.
- Filtering, aggregation and Fusion data from multiple readers in real time, Detect anomalies.
- Analysis and research related data in external databases.
- Etc.

These advantages make the application more robust by having two major concepts applied for the traceability of luggage. First, it provides a flexible and convenient structure for monitoring and management of luggage. Second, it can be adapted to the needs of airlines and passengers for an organization more personalized in the airport.

5 Conclusion

During the last decade, this wide spectrum of applications that have been observed is nothing but the result of the emergence of research and development, which were conducted in different areas such as logistics, air, and military to field security and personal services. It falls into the so-called Internet objects or connected objects. Connected objects and communicating could provide opportunities for the industry in the years to come, and a new middleware architecture become strategic regarding a massive rise of "unintelligent" internet of objects is expected by 2020. This work response to this need with a vision of connecting functions placed in objects.

An estimated 50 billion items could be tagged by 2020. This ability of communication will develop virtual reality devices. Thus research and innovation in the field of data integration collected by technologies that deal with communication between objects is still needed to support this development and is a major factor for the success of these technologies and their relevance.

This paper introduces RFID technology and the role of RFID middleware in the enterprise information systems. It also describes the design of our middleware MedRFID and its architecture. With such an innovative architecture, MedRFID provides extensibility, scalability, abstraction and ease of deployment like the FindLuggage application demonstrates. To be compatible with different areas including logistics and airports, it integrates respectively EPCglobal standards and IATA standards making it easily deployed in heterogeneous sites.

References

1. Ajana, M. E., Boulmalf, M., Harroud, H. & Hamam, H. (2009). A Policy Based Event Management Middleware for Implementing RFID Applications, Proceedings of WiMOB 2009 5th International Conference on Wireless and Mobile Computing, Networking and Communications, ISBN 978-0-7695-3841 -9, Marrakesh, Morocco, October 12-14, 2009.
2. Al-Mousawi, H. (2004). Performance and Reliability of Radio Frequency Identification (RFID).
3. Burnell, J. (2008). What Is RFID Middleware and Where Is It Needed?, In: RFID Update.
4. Floerkemeier, C., Roduner, C. & Lampe, M. (2007). RFID Application Development with the Accada middleware Platform. IEEE Systems Journal, Vol.1 No.2, pp. 82-94, ISSN 1932- 8184.
5. Floerkemeier, C. & Lampe, M. (2005). RFID Middleware Design: Addressing Application Requirements and RFID Constraints, Proceedings of SOC'2005 Smart Objects Conference, pp. 219-224, ISBN 1 -59593-304-2, Grenoble, France, October, 2005.
6. Glasser, D. J., Goodman, K. W. & Einspruch, N. G. (2007). Chips, Tags and Scanners: Ethical Challenges for Radio Frequency Identification. Ethics and Information Technology, Vol.9, No.2, pp. 101 -109, ISSN 1388-1957.
7. Molnar, D. & Wagner, D. (2004). Privacy and Security in Library RFID: Issues, Practices, and Architectures, Proceedings of ACM CCS 2004 11th Conference on Computer and Communication Security, ISBN 1 -58113- 961 -6, Washington, DC, USA, October, 2004.
8. Prabhu, B. S., Su, X., Ramamurthy, H., Chu, C. & Gadh, R. (2005 a). WinRFID: A Middleware for the Enablement of Radio Frequency Identification (RFID) Based Applications, In: Wireless Internet for the Mobile Enterprise Consortium (WINMEC).
9. Prabhu, B. S., Su, X., Ramamurthy, H., Chu, P.,Qiu, C. & Gadh, R. (2005 b). WinRFID: Middleware for Distributed RFID Infrastructure, In: Wireless Internet for the Mobile Enterprise Consortium (WINMEC).
10. Sheng, Q. Z., Li, X. & Zeadally, S. (2008). Enabling Next-Generation RFID Applications: Solutions and

Challenges. IEEE Computer, Vol.41, No.9, pp. 21 - 28, ISSN 0018-9162.
11. United States Government Accountability Office (2005). Information Security Radio Frequency Identification Technology in the Federal Government.
12. A.Raghib, B.Abou El Majd (2014)."Multi-level optimal deployment of RFID readers by using particle swarm optimization", Proceedings of META'14 5th International Conference on Metaheuristics and Nature Inspired Computing, Marrakesh, Morocco, October 27-31, 2014.
13. Indrajit Bhattacharya, Uttam Kumar Roy, "Optimal Placement of Readers in an RFID Network Using Particle Swarm Optimization," International Journal of Computer Networks & Communications– IJCNS, vol.2, no.6, November 2010.
14. E. Jaselskis, T.El-Misalami, Implementing radio frequency identification in the construction process, Journal of Construction Engineering and Management (ASCE) 129 (6) (2003) 680–688.
15. S. Chae, T. Yoshida, Application of RFID Technology to prevention of collision accident with heavy equipment, Journal of Automation in Construction 19 (2010) 368–374.
16. C. Legner, F. Thiesse, RFID-Based maintenance at Frankfurt airport, IEEE Pervasive Computing 5 (1) (2006) 34–39.
17. R. Want. An introduction to RFID technology. IEEE Pervasive
Computing, 5(1):25–33, Jan.-March 2006.
18. G. M. Gaukler, "Item-level RFID in a retail supply chain with stock-out based substitution," IEEE Transactions on Industrial Informatics, vol. 7, no. 2, pp. 362-370, 2011.
19. T.-M. Choi, "Coordination and risk analysis of VMI supply chains with RFID technology," IEEE Transactions on Industrial Informatics, vol. 7, no. 3, pp. 497-504, 2011.
20. J. Vales-Alonso, V. Bueno-Delgado, E. Egea-Lopez, F. J. Gonzalez Castano, and J. Alcaraz, "Multiframe maximum-likelihood tag estimation for RFID anticollision protocols," IEEE Transactions on Industrial Informatics, vol. 7, no. 3, pp. 487-496, 2011.
21. Y.-H. Chen, S.-J. Horng, R.-S. Run, J.-L. Lai, R.-J. Chen, W.-C. Chen, Y. Pan, and T. Takao, "A novel anti-collision algorithm in RFID systems for identifying passive tags," IEEE Transactions on Industrial Informatics, vol. 6, no. 1, pp.105-121, 2010.
22. Roy W., Daniel M.R., "Ubiquitous Electronic Tagging," IEEE Distributed Systems Online, Vol. 1, No 2, Jan. 2004.

Hybrid resolution approaches for dynamic assignment problem of reusable containers

Mohammed Rida Ech-Charrat [1,*], Khalid AMECHNOUE [1] and Tarik Zouadi [2]

[1] *Laboratory of Innovative Technologies, National School of Applied Sciences, Tangier, Morocco*
[2] *BEAR Lab, Rabat Business School, International University of Rabat (UIR), Rabat, Morocco*

Abstract. In this study, we are interested in the reusing activities of reverse logistics. We focus on the dynamic assignment of reusable containers problem (e.g. gas bottles, beverages, pallets, maritime containers, etc.). The objective is to minimize the collect, reloading, storage and redistribution operations costs over a fixed planning horizon taking into account the greenhouse gas emissions. We present a new generic Mixed Integer Programming (MIP) model for the problem. The proposed model was solved using the IBM ILOG CPLEX optimization software; this method yield exact solutions, but it is very time consuming. So we adapted two hybrid approaches using a genetic algorithm to solve the problem at a reduced time (The second hybrid approach is enhanced with a local search procedure based on the Variable Neighborhood Search VNS). The numerical results show that both developed hybrid approaches generate high-quality solutions in a moderate computational time, especially the second hybrid method.

1 Introduction

To meet environmental and economic requirements, supply chain management (SCM) is no longer limited to direct flow management but it includes also the returns flow problems. Reverse logistics is a quite vast field, since the returned products appear in many types of recovery such as remanufacturing, recycling, reusing… (Hanafi et al., [1]). In this study, we are interested in the reusing activity (using a product more than once). The product concerned by our research is the reusable container due to its economic and environmental importance (Bhattacharjya et al. [2]).

The reusable containers require normally certain treatments before they are redistributed, thus reusing containers problem addressed in our research is not limited only to a simple recovery and transportation; it is a logistic system that involves also the cleaning and reloading of containers, as well as their storage and redistribution.

Producers in several countries are facing increasing market pressures to use ecological packaging. Castillo et al. [3] formulate an optimal configuration for the reusable bottle production and distribution activities of a large soft drink manufacturer located in Mexico City. Goudenege et al. [4] propose a generic model for reverse logistics management focused on reusable containers. The authors adapt the model to the specific requirements of companies. Atamer et al. [5] present a study that emphasizes on pricing and production decisions while treating reusable containers with stochastic customer demand. Accorsi et al. [6] propose an original conceptual framework for the integrated design of a food packaging and distribution network. The paper compares a multi-use system to a traditional single-use packaging in order to quantify the economic returns and the environmental impacts of the reusable plastic container (RPC). In the same sense, evaluating the environmental impact is one of the aims of the green supply chain. The signed Kyoto protocol engage many countries and companies as a consequence to a carbon emission quota. Several methodologies are used to calculate carbon emissions (Retel Helmrich et al. [7]). Absi et al. [8] propose four types of carbon emission constraints (Periodic, cumulative, global and rolling) for the multi-sourcing lot-sizing problem, which can be used and adapted to several cases. In this study, we propose a generic mathematical model that helps companies minimizing transportation, storage, containers reloading and investment costs. The model proposes a flow management concerning the reusable containers between the warehouses and the stores. The full delivered containers are consumed and recuperated empty thereafter by the warehouses to be reloaded and redistributed again to the stores on a fixed planning horizon. The mathematical model takes into consideration the carbon emission constraint during the distribution and collection activities.

* Corresponding author: charrat.mohammed@gmail.com

2 Mathematical model:

$$\text{TransC}=\sum_{i=1}^{N}\sum_{j=1}^{M}\sum_{t=1}^{T}\left(\left(Z_{i,j,t}*\text{pur}_{i,j}\right)+\left(Y_{i,j,t}*\text{recov}_{i,j}\right)+\begin{pmatrix}\left(XF_{i,j,t}+XE_{i,j,t}\right)\\ *\text{trans}_{i,j}*\text{dist}_{i,j}\end{pmatrix}\right)$$

$$\text{HoldC}=\sum_{i=1}^{N}\sum_{t=1}^{T}\left(\left(IFW_{i,t}*h_{f_{w_i}}\right)+\left(IEW_{i,t}*h_{e_{w_i}}\right)\right)+$$
$$\sum_{j=1}^{M}\sum_{t=1}^{T}\left(\left(IFC_{j,t}*h_{f_{c_j}}\right)+\left(IEC_{j,t}*h_{e_{c_j}}\right)\right)+\sum_{i=1}^{N}\begin{pmatrix}\left(IOEW_i*h_{e_{w_i}}\right)+\\ \left(IOFW_i*h_{f_{w_i}}\right)\end{pmatrix}$$

$$\text{ProdC} = \sum_{i=1}^{N}\sum_{t=1}^{T}\left(Q_{i,t}*\text{Prod\_cost}_i\right)$$

The integrated planning model is shown in the following:

$$\text{Min}\ (\text{TransC}+\text{HoldC}+\text{ProdC}) \tag{1}$$

Subject to:

$$IFW_{i,1} = IOFW_i + Q_{i,1-Delay} - \sum_{j=1}^{M}\left(XF_{i,j,1}\right)\ \forall i, (\forall Delay < 1 \textit{ if not } Q = 0) \tag{2}$$

$$IFW_{i,t} = IFW_{i,t-1} + Q_{i,t-Delay} - \sum_{j=1}^{M}\left(XF_{i,j,t}\right)\ \forall i, (\forall t > Delay, \text{if not } Q = 0) \tag{3}$$

$$IEW_{i,1} = IOEW_i - Q_{i,1} + \sum_{j=1}^{M}\left(XE_{i,j,1}\right)\ \forall i \tag{4}$$

$$IEW_{i,t} = IEW_{i,t-1} - Q_{i,t} + \sum_{j=1}^{M}\left(XE_{i,j,t}\right)\ \forall i \tag{5}$$

$$IFC_{j,1} = IOFC_j - dem_{j,1} + \sum_{i=1}^{N}\left(XF_{i,j,1-d_{j,i}}\right)\ \forall j, (\forall d_{j,i} < 1 \textit{ ifnot } XF = 0) \tag{6}$$

$$IFC_{j,t} = IFC_{j,t-1} + \sum_{i=1}^{N}\left(XF_{i,j,t-d_{j,i}}\right) - dem_{j,t}\ \ \forall j, \forall t > 1, (\forall d_{j,i} < t \text{ else } XF = 0) \tag{7}$$

$$IEC_{j,1} = IOEC_j - \sum_{i=1}^{N}\left(XE_{i,j,1+d_{j,i}}\right)\ \forall j, (\forall d_{j,i} < T \text{ else } XE = 0) \tag{8}$$

$$IEC_{j,t} = IEC_{j,t-1} - \sum_{i=1}^{N}\left(XE_{i,j,t+d_{j,i}}\right) + dem_{j,t-1}\ \ \forall j,\ \forall t > 1, \left(\forall d_{j,i} \le T - t \text{ else } XE = 0\right) \tag{9}$$

$$\sum_{i=1}^{N}\sum_{j=1}^{M}\sum_{t=1}^{T}\left(\left(XF_{ijt} + XE_{i,j,t}\right) * CO2 * dist_{i,j}\right) \le Emax * \sum_{j=1}^{M}\sum_{t=1}^{T} dem_{j,t} \tag{10}$$

$$XF_{ijt} \le H * Z_{i,j,t}\ \ \forall i, \forall j,\ \forall t \tag{11}$$

$$XE_{ijt} \le H * Y_{i,j,t}\ \forall i, \forall j, \forall t \tag{12}$$

$$IFW_{it} + IEW_{it} \le s\_c\_w_i\ \forall i,\ \forall t \tag{13}$$

$$IFC_{jt} + IEC_{jt} \le s\_c\_c_j\ \forall j,\ \forall t \tag{14}$$

$$XE_{ijt}, XF_{ijt}, IFW_{it}, IEW_{it}, IFC_{jt}, IEC_{jt}, IOFW_i, IOEW_i, Q_{i,t} \ge 0 \tag{15}$$

The objective function (1) calculate the solution fitness. It minimizes the transportation, reloading, investment and holding cost of reusable containers between warehouses and stores. Constraints (2& 3) and (4 & 5) are respectively the inventory flow conservation equations for full and empty reusable containers in the warehouses. Constraints (6 & 7) and (8 & 9) are respectively the inventory flow conservation equations for full and empty reusable containers in the stores. Constraint (10) presents the unitary carbon emission over the whole horizon that cannot be larger than the maximum unitary environmental impact allowed. Constraints (11) and (12) guarantee respectively the cancellation of full and empty reusable containers deliveries when no delivery is programmed. Constraints (13) and (14) guarantee respectively the respect of the inventory capacity of the warehouses and stores.

3 Solving approaches

3.1 Hybrid algorithm 1 (HA1)

In this section, we propose an approach made of hybridization between an evolutionary algorithm (the genetic algorithm) and an exact resolution (the mixed integer programming) to solve the problem in a moderate computational time. Some interesting applications of GA are presented by Supithak et al. [9], Rezaei et al. [10], and Zouadi et al. [11].

3.1.1 Solution encoding

The proposed binary encoding in Figure 1 presents 3-dimensional vectors which are the period of the planning horizon (T), the warehouses number (N) and the double of the stores number (2×M), which constitutes the collect and the distribution parts. In the first part of the vector, we find the reusable containers deliveries binary decision variables between the warehouses and the stores over the planning horizon. While in the second part of the vector, we have the reusable containers collection binary decision variables.

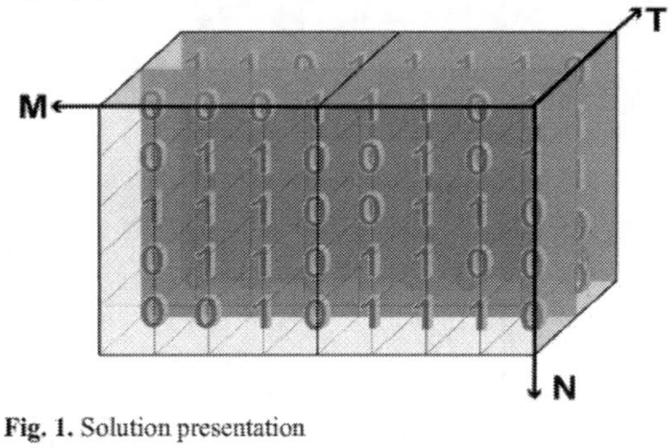

Fig. 1. Solution presentation

3.1.2 Crossover

Following the nature of the solutions encoding, we proposes a new crossover operator to cross the two different 3-dimensional offsprings. The proposed crossover is presented in the Figure 2.

The proposed crossover operator consists of splitting the 3-dimensional vectors into T matrix (2-dimensional vectors), in a way that each matrix presents the full reusable containers deliveries and the empty reusable containers collection binary decisions over a period *t*. The crossover operator consists of crossing randomly two matrixes issued from two parents using one of the crossing operators proposed by Toledo et al. [12].

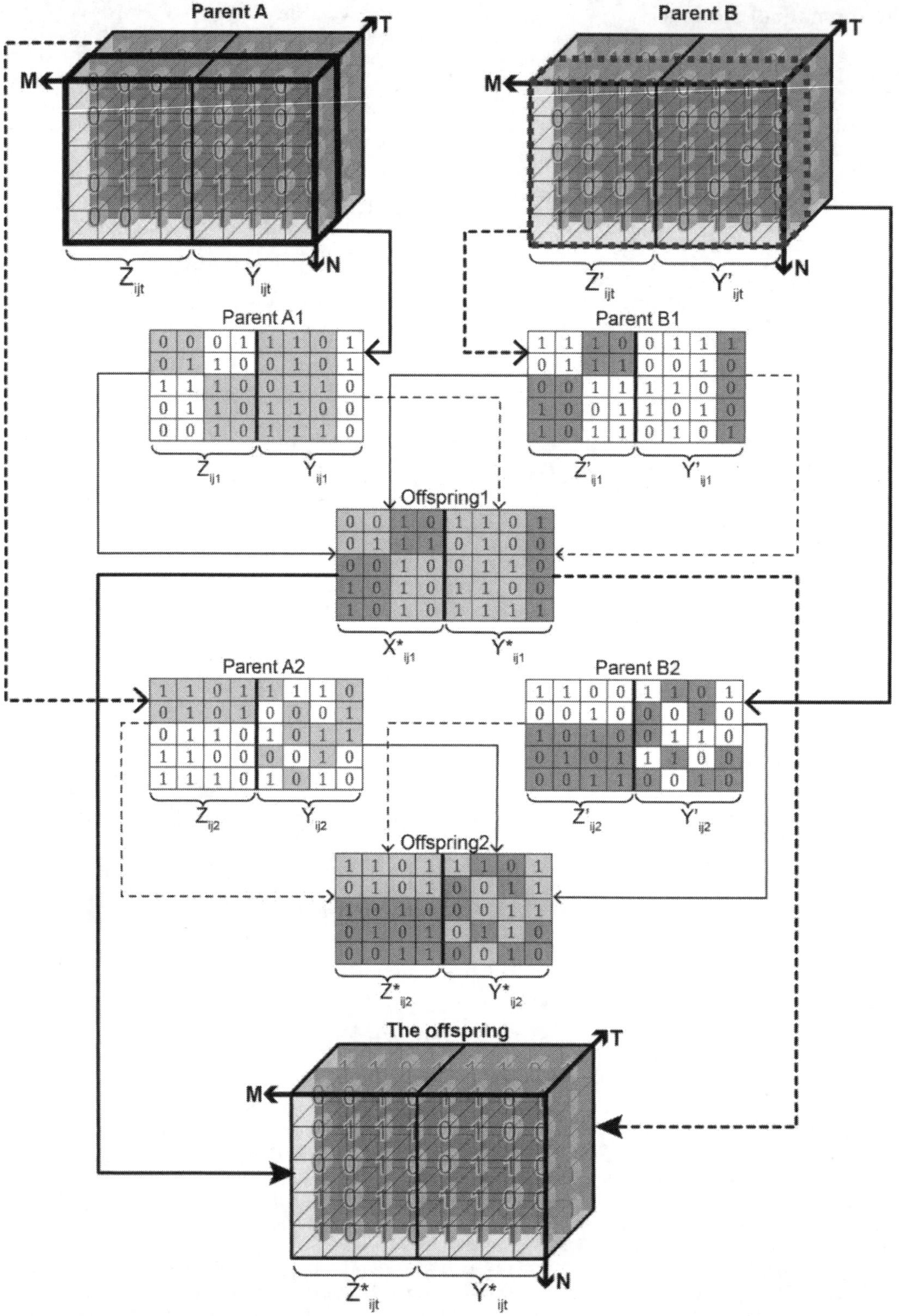

Fig. 2. Crossover operator

In the example presented in the Figure 2, all of the two 3-dimensional vectors (Parent A, Parent B) are divided into two matrixes {(PA1, PA2) ;(PB1, PB2)} representing the deliveries and collection decision variables of the two period of the planning horizon (T). The crossover operator consists of crossing the first half of the matrix PA1 (presenting the deliveries decision variables) with the first half of the matrix PB1, and the second half of the PA1 (presenting the collection decision variables) with the second half of the PB1 to get the offspring1 presenting the first period matrix decision crossing. The same procedure is launched to cross the PA2 and PB2 to get the second offspring. The two offspring 1 and 2 are assembled in a 3-dimensional vector presenting the resulted offspring.

3.1.3 Mutation

In this paper, we use the implementation of four mutation operators proposed by Teledo et al. [12]. The first operator resides on changing the value of a randomly chosen variable. In the second operator, two variables are randomly chosen from the same column and their values are exchanged. In the third operator, two variables are randomly chosen from the same column and their values are exchanged. The last operator consists of applying the first mutation operator twice. We select randomly one of these four mutations.

3.1.4 Exact resolution to determine collected and delivered quantities

At each iteration of the Hybrid algorithm, the generated offspring determine the solution for only the binary decision variables of full and empty reusable containers deliveries or collects. These propositions generated by the offspring are integrated in the mathematical model which is solved by IBM ILOG CPLEX platform. The solution returned by CPLEX defines optimal reusable containers quantities to deliver and to collect according to the binary decisions proposed by the offspring structure resulting from the crossover and mutation operators.

3.2 Hybrid algorithm 2 (HA2)

In the second hybrid structure (HA2), we consider the same hybridization architecture with the use of a memetic algorithm instead of the genetic used in the (HA1).

The proposed memetic algorithm keeps the same genetic algorithm architecture (The same solution presentation, crossover and mutation operators, and the exact resolution). In addition to genetic features, the memetic algorithm uses a local search to enhance the solutions quality. We use the Variable Neighborhood Search (VNS) approach; The VNS concept is to exploit systematically the idea of neighborhood change (intensification). (Hansen et al. [13])

The VNS in this configuration exploits three different local searches. The metaheuristic logic is to explore distant neighborhoods of the current solution (offspring) until a stopping criterion, and moves from solution to another and from a local search procedure to another only if an improvement was made. In the case where no improvement is made, the metaheuristic pass to the second local search exploitation with the best-found solution found until a stopping criterion. If no improvement is made, the same strategy is followed by launching the third local search. In the case of finding a better solution in the second or in the third local search, the VNS starts from the beginning by exploring the first local search using the best solution found. We propose three local searches. The first local search consists of permuting the matrixes presenting the full and empty reusable containers deliveries and collets decisions on each period T_i. The move A in Figure 3 shows an example of the used local search. The second local search consists of permuting the lines of the matrix T_i presenting the full and empty reusable containers deliveries and collets decisions of the period *i*. The move B in Figure 3 shows an example of the used local search. The third local search consists on inverting the bit's value, one by one of each line in the matrix T_i until a stopping criterion. The move C in Figure 3 shows an example of the used local search.

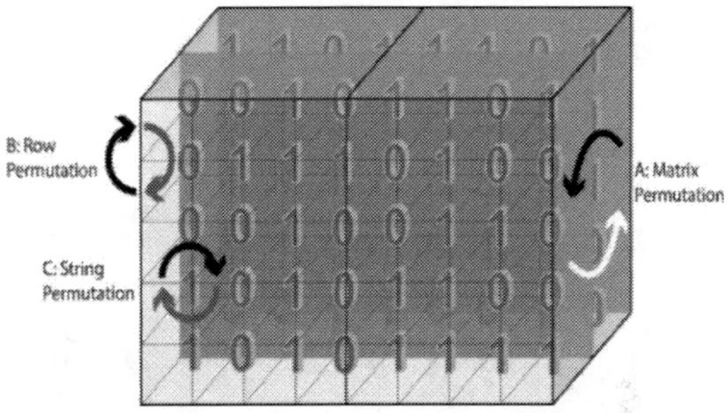

Fig. 3. The used local search procedures

4 Result and discussion

To prove the proposed algorithms performance, we summarize in this section the numerical experiments performed on these approaches. The obtained results by the first Hybrid method 1 (HA1), and the second Hybrid method (HA2) are compared with those obtained by CPLEX. The instances are derived from the article of Teunter et al. [14], Zouadi et al. [11] and Absi et al. [8]. Five different types of demand and return patterns (stationary, linearly increasing, linearly decreasing, seasonal peak in the middle and seasonal valley in the middle, 6 horizon lengths, 4 set of warehouses, and 8 sets of clients are considered.

Table 1. The results obtained according to the number of periods.

Periods	Cplex		HA1				HA2			
	AS	AT(s)	AS	GHA1C	AT(s)	MaxG	AS	GHA2C	AT(s)	MaxG
5	80889,34	66	81989,44	1,36%	73	7,00%	81876,19	1,22%	75	3,45%
10	193663,88	332	198931,54	2,72%	211	17,27%	198428,01	2,46%	233	5,76%
15	315144,58	3425	324977,09	3,12%	396	8,10%	321857,16	2,13%	424	8,10%
20	433879,89	8409	447156,61	3,06%	501	13,38%	443078,14	2,12%	779	11,91%
25	577111,09	13353	592923,93	2,74%	779	9,10%	586864,27	1,69%	1054	9,10%
30	747121,18	24438	764529,10	2,33%	968	11,80%	757431,45	1,38%	1634	9,80%

Compared to the results obtained by CPLEX, HA2 gives solutions that are close by 1.83% to the optimal solutions while the HA1 is less efficient, it gives solutions that are close by 2,56 % to the optimal solutions. The computational time of the HA2 is greater than the HA1 due to the used local search.

The quality of solutions depends on the length of the planning horizon. For small horizons (5 periods), both methods give solutions that are close to the optimal solutions, especially HA2, and in particular when the number of client and warehouses is small. While for long horizons, the solutions are pretty far from being close to optimal. Regarding the computational time, CPLEX is rather efficient to prove optimality for small size instances, but on larger ones, the computational time becomes more important and exceeds half an hour on many instances without finding an optimal solution.

5 Conclusion

This contribution proposes a new generic mixed integer programming model of dynamic assignment of reusable containers, and two hybrid approaches to solve the problem in a moderate computational time. The two developed hybrid approaches generate high-quality solutions. This addition is a base to develop more integrated models by considering several constraints such as the number of products, multi-level structures and cooperation between different actors.

References

1. J. Hanafi, S. Kara, H. Kaebernick, Reverse logistics strategies for end-of-life products, The International Journal of Logistics Management, Vol. 19 Iss: 3, pp.367 - 388, (2008).
2. J. Bhattacharjya, P. Kleine-Moellhoff, Environmental Concerns in the Design and Management of Pallets, International Federation for Information Processing, 569–576, (2013).
3. E. Castillo, J.K. Cochran, Optimal Short Horizon Distribution Operations in Reusable Container Systems, the Journal of the Operational Research Society, Vol. 47, No. 1, pp. 48-60, (1996).
4. G. Goudenege, C. Chu, Z. Jemai, Reusable Containers Management: From a Generic Model to an Industrial Case Study, Supply Chain Forum (An International Journal), Vol. 14, No. 2, (2013).
5. B. Atamer, İ.S. Bakal, Z.P. Bayindir, Optimal pricing and production decisions in utilizing reusable containers, International Journal of Production Economics, V 143, Issue 2, Pages 222–232, (2013).
6. R. Accorsi, A. Cascini, S. Cholette, R. Manzini, Mora, C., Economic and environmental assessment of reusable plastic containers: A food catering supply chain case study, International Journal of Production Economics, Volume 152, June 2014, Pages 88–101, (2014).
7. M. J. Retel Helmrich, R. Jans, W. V. D. Heuvel, A. P. M. Wagelmans, The economic lot-sizing problem with an emission capacity Constraint. European Journal of Operational Research, Volume 241, Issue 1, Pages 50–62, (2015).
8. N. Absi, S. Dauzère-Pérès, S. Kedad-Sidhoum, B. Penz, C. Rapine, Lot sizing with carbon emission constraints. European Journal of Operational Research, Vol. 227 (1), Pages 55–61, (2013).
9. W. Supithak, S. D. Liman, E. J. Montes, Lot-sizing and scheduling problem with earliness tardiness and setup penalties, Computers & Industrial Engineering, Volume 58, Issue 3, 363–372, (2010).
10. Rezaei, J. and Davoodi, M., Multi-objective models for lot-sizing with supplier selection, International Journal of Production Economics, Volume 130, Issue 1, Pages 77–86, (2011).
11. T. Zouadi, A. Yalaoui, M. Reghioui, K. E. EL Kadiri, K. E., Lot-sizing for production planning in a recovery system with returns, RAIRO Operations Research, Vol 49 1, 123-142, (2015).
12. C. F. M. Toledo, R. R. Oliveira, P. M., A França, hybrid multi-population genetic algorithm applied to solve the multi-level capacitated lot sizing problem with backlogging, Computers & Operations Research,Volume 40, Issue 4, Pages 910–919, 2013.
13. P. Hansen, N. Mladenovic, D. Perez-Brito, Variable neighborhood decomposition search, Journal of Heuristics, Vol. 7, Issue 4, pp 335–350, (2001).
14. R.H. Teunter, Z. P. Bayindirand, W. V. D. Heuvel, W., International Journal of Production Research, Volume 44, Issue 20, 4377-4400, (2006).

The Mean-CVaR Model for Portfolio Optimization Using a Multi-Objective Approach and the Kalai-Smorodinsky Solution

R. Aboulaich[1], *R.* Ellaia[1], *S.* Elmoumen[2], *A.* Habbal[3], and *N.* Moussaid[4,*]

[1]*LERMA, Mohammadia School of Engineering, Mohammed V University Rabat, Morocco*
[2]*LIMSAD, Mathematics and computing Department, Ain Chock Sciences Faculty, Casablanca, Morocco*
[3]*INRIA, University of Nice Sophia-Antipolis, France*
[4]*LMA, Hassan II University Casablanca, FST, Mohammedia, Morocco*

Abstract. The purpose of this work is to present a model for portfolio multi-optimization, in which distributions are compared on the basis of tow statistics: the expected value and the Conditional Value-at-Risk (CVaR), to solve such a problem many authors have developed several algorithms, in this work we propose to find the efficient boundary by using the Normal Boundary Intersection approach (NBI) based on our proposed hybrid method SASP, since the considered problem is multi-objective, then we find the Kalai-smorodinsky solution.

1 Introduction

Mean-risk models are still the most widely used approach in the practice of portfolio selection. With mean-risk models, return distributions are characterized and compared using two statistics: the expected value and the value of a risk measure. Thus, mean-risk models have a ready interpretation of results and in most cases are convenient from a computational point of view. On the other hand in the practice of describing a distribution by just two parameters involves great loss of information.

It is well known that the risk measure plays an important role in making the decisions. Variance was the first risk measure used in mean-risk models (Markowitz 1952) and, in many proposals of new risk measures (see for example Fishburn (1977), Yitzhaki (1982), Konno and Yamazaki (1991), Ogryczak and Ruszczynski (1999, 2001), Rockafellar and Uryasev (2000, 2002)). Another risk measure, the Conditional Value-at-Risk (CVaR))[16],[20], is growing in popularity. CVaR has attractive theoretical properties: it controls the magnitude of losses beyond Value-at-Risk (VaR) and it is coherent (see for example Artzner 1999, Acerbi and Tasche 2002, Tasche 2002, Pflug 2000, Rockafellar and Uryasev 2002). In this paper we propose to solve the problem of portfolio selection, which is a multi-objective problem, first by using the NBI approach [8] based on SASP method [7], implemented in Matlab to find the efficient boundary, after we use the game theory [3, 4][9][2][1] to determine the Kalai-smorodinsky solution.

*e-mail: moussaid_noureddine@yahoo.fr

2 The portfolio selection problem

The problem of portfolio selection with one investment period is an example of the general problem of deciding between random variables when larger outcomes are preferred. Decisions are required on the amount (proportion) of capital to be invested in each of a number of available assets such that at the end of the investment period the return is as high as possible. Consider a set of n assets, with asset j in $\{1, ..., n\}$ giving a return R_j at the end of the investment period. R_j is a random variable, since the future price of the asset is not known. Let x_j be the proportion of capital invested in asset j ($x_j = w_j/w$ where w_j is the capital invested in asset j and w is the total amount of capital to be invested), and let $x = (x_1, ..., x_n)$ represent the portfolio resulting from this choice. This portfolio's return is the random variable: $R_x = x_1R_1 + ... + x_nR_n$, with distribution function $F(r) = P(R_x \leq r)$ that depends on the choice $x = (x_1, ..., x_n)$.

To represent a portfolio, the weights $(x_1, ..., x_n)$ must satisfy a set of constraints that forms a feasible set $\mathbf{A}$ of decision vectors. The simplest way to define a feasible set is by the requirement that the weights must sum to 1 and short selling is not allowed. For this basic version of the problem, the set of feasible decision vectors is:

$$\mathbf{A} = \{(x_1, ..., x_n)/ \sum_{i=1}^{n} x_j = 1, x_j \geq 0, \forall j \in \{1, ..., n\}\}$$

If we consider a different portfolio defined by the decision vector $y = (y_1, ..., y_n) \in \mathbf{A}$, where y_j is the proportion of capital invested in asset j. The return of this portfolio is given by the random variable $R_y = y_1R_1 + ... + y_nR_n$.

The problem of choosing between portfolio $x = (x_1, ..., x_n)$ and portfolio $y = (y_1, ..., y_n)$ becomes the problem of choosing between random variables R_x and R_y. The criteria by which one random variable is considered "better" than another random variable need to be specified and models for choosing between random variables (models for preference) are required. a lot of models are proposed, the markowitz model, the mean model and the mean variance model.

3 Conditional Value-at-Risk

Let R_x be a random variable representing the return of a portfolio x over a given holding period and $\alpha \in (0,1)$ a percentage which represents a sample of "worst cases" for the outcomes of R_x.
The definition of CVaR at the specified level α is the mathematical transcription of the concept "average of losses in the worst α of cases" (Acerbi and Tasche 2002), where a "loss" is a negative outcome of R_x (thus the loss associated with R_x is described by the random variable -R_x). An important result, proved by Rockafellar and Uryasev (2000, 2002), and independently by Ogryczak and Ruszczynski (2002), is that the CVaR of a random variable R_x can be calculated by solving an optimization problem. Moreover, CVaR can be minimized over the set of feasible decision vectors. These results are summarized below:
Let R_x be a random variable depending on a decision vector x that belongs to a feasible set **A**, and $\alpha \in (0,1)$. Consider the function:

$$F_\alpha(x,\upsilon) = \frac{1}{\alpha}E\{[-R_x + \upsilon]^+\} - \upsilon$$

Then:
(a) As a function of υ, F_α is finite and continuous (hence convex) and

$$CVaR_\alpha(R_x) = \min_{\upsilon \in R} F_\alpha(x,\upsilon).$$

(b) Minimizing $CVaR_\alpha$ with respect to $x \in A$ is equivalent to minimizing F_α with respect to $(x,\upsilon) \in A \times \boldsymbol{R}$:

$$\min_{x\in A} CVaR_\alpha(R_x) = \min_{(x,\upsilon)\in A\times R} F_\alpha(x,\upsilon). \quad (1)$$

(c) $CVaR_\alpha(R_x)$ is convex with respect to x and $F_\alpha(x,\upsilon)$ is convex with respect to (x,υ).

The Mean-CVaR model
Let E be the expected value of R_x, the Mean-CVaR model can be formulated for the portfolio selection problem as follows:

$$\min_{x\in A} F = (f_1, f_2) \quad (2)$$

Where $f_1 = CVaR$ and $f_2 = -E$
Since 1, the problem 2 becomes:

$$\min_{x\in A}(F_\alpha(x,\upsilon), -E) \quad (3)$$

4 Normal Boundary Intersection

Let x_i^* and f_i^* denote respectively the minimizer and minimum value of the f_i and let F^* denote the shadow minimum, i.e., the vector whose components are f_i^*. Consider the shifted pay-off matrix Φ whose i^th column is $F(x_i^*)-F^*$. The Convex Hull of Individual Minima or CHIM is defined as the set of points that are convex combinations of the columns of Φ , i.e., $\{\Phi\beta : \beta_i \geq 0, \sum_i \beta_i = 1\}$.
For a two dimensional problem illustrated in Figure 1, CHIM is represented by segment AB.

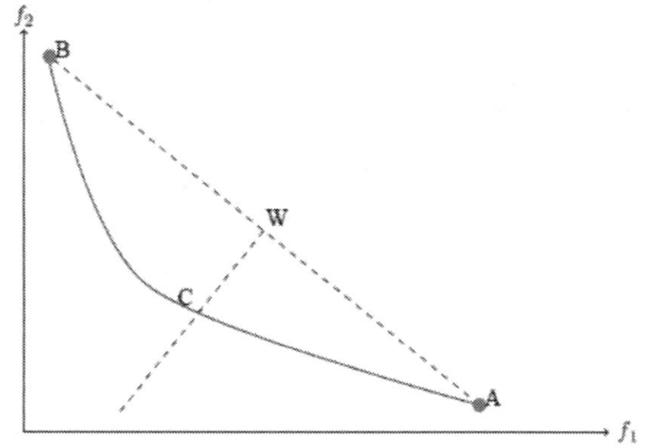

Figure 1. An illustrative integrated design

The idea behind NBI is to pick an even spread of points on the CHIM (for example W in Fig.1), and find the intersection point between the efficient frontier and a set of parallel normals emanating from the chosen set of points on the CHIM (C in Fig.1). Given a convex combination parameter vector β, and a normal direction n pointing towards the origin, the point of intersection between the normal emanating from $\Phi\beta$ and the efficient frontier can be found by solving the following NBI_β subproblem :

$$\begin{cases} \min_{x\in A} t; \\ \Phi\beta + tn = F(x) - F^*. \end{cases} \quad (4)$$

By solving subproblem NBI_β 4 for different settings of β, various points on the efficient frontier can be generated. The advantage of the β parameter is that an even spread of β parameters corresponds to an even sped of points on the CHIM.

5 Kalai Smorodinsky solution

In this paper, we research the Kalai Smorodinsky solution [18], witch is the intersection between the Pareto front and the line passing through the utopian point Ut and the disagreement point D. Note,

$$Ut = \begin{pmatrix} f_1(x^*) \\ f_2(y^*) \end{pmatrix}, D = \begin{pmatrix} f_1(y^*) \\ f_2(x^*) \end{pmatrix} \text{ and } \tau = \frac{Ut - D}{\| Ut - D \|}$$

where,

$$\begin{cases} x^* = Arg \min\limits_{x} f_1(x) \\ y^* = Arg \min\limits_{y} f_2(y). \end{cases} \quad (5)$$

Theorem 1 *A solution is symmetric, Pareto optimal[9][2][1], invariant under affine transformations of utility scale, and individually monotonic if and only if it is the Kalai-Smorodinsky solution [18].*

To search the Kalai Smorodinsky solution [5], we proposed to solve the following problem:

$$(KS)\begin{cases} \max\limits_{y,t} t, \\ s.c\ F(x) = D + t\tau, \end{cases} \quad (6)$$

To solve such problem we propose to use the SASP method [7].

6 Numerical Example

6.1 An application to portfolio optimization

Let (SAMIR, ATWB, AGM, LAM, GAZ, IAM, ONA) be the assets, quoted in the Casablanca Stock Exchange, with history of 5 years (see the following Figure).

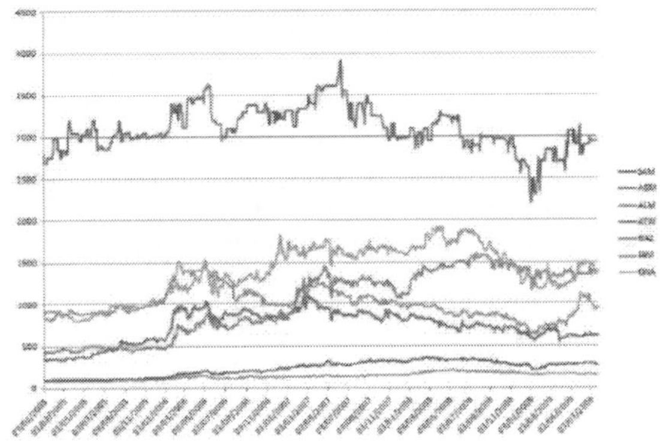

Figure 2. An illustrative integrated design

First we are going to solve the problems (8) and (9):

$$\min_{x \in A} CVaR \quad (7)$$

and

$$\min x \in A - E \quad (8)$$

Then we have to find the efficient frontier with the NBI method based on the SASP method, for this, we have to solve the problems below for deferent β with $\beta \in \{\beta \in IR^n, \sum_{i=1}^{n} \beta_i = 1, \beta_i \geq 0\}$:

$$\begin{aligned} &\max \quad t \\ s.t.\ \ &\Phi\beta + tn = F(x) - F^* \\ &x \in A \end{aligned}$$

The Kalai Smorodinsky solution can be find by solving the following problem:

$$(KS)\begin{cases} \max\limits_{x,t} t, \\ s.c\ F(x) = D + t\tau, \end{cases} \quad (9)$$

6.2 Results

The results are provided by the following figure:

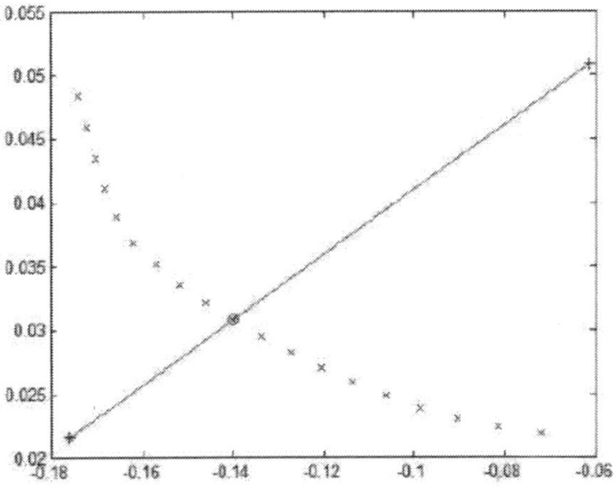

Figure 3. An illustrative integrated design

As we can see clearly the NBI parametrization of the efficient front naturally bears the property of even spread, and the Kalai Smorodinsky solution is closest to the ideal solution (maximum E, minimum CVaR), the Kalai Smorodinsky portfolio "PF" (36% in ATW, 37% in GAZ and 27% in IAM) history is given with the other histories in the following figure:

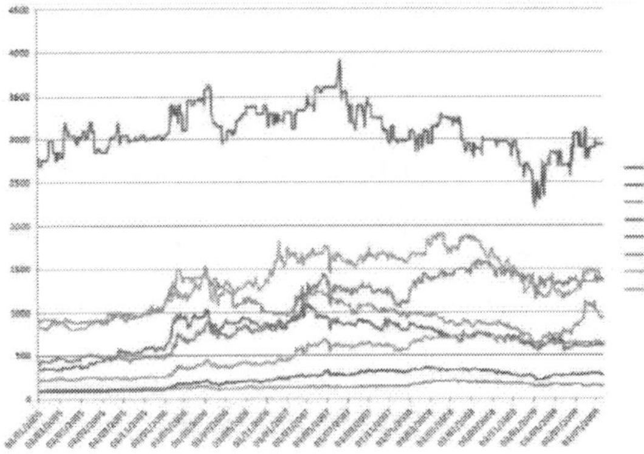

Figure 4. An illustrative integrated design

If we choose another optimal portfolio, for example the portfolio corresponding to (0,1061, 0,0248) we obtain (13,81% in AGM,5,02% ALM, 22,93% ATW, 25,5% GAZ and 32,75 IAM) and for (0,1722 , 0,0459) we obtain (14,21% ATW, 85,79% GAZ), we can't have an equilibrium between maximum E and minimum CVaR.

7 Conclusion

In this work, we proposed a solution of bi-objective model in order to obtain the optimal portfolio. First we used the Normal Boundary Intersection approach (NBI) to compute the efficient front, and then the optimal solution is given by Kalai Smorodinsky solution. The obtained results show the efficiency of the proposed approach.

References

[1] B. Abou El Majd, J.A. Désidéri, and A. Habbal. *Aerodynamic and structural optimization of a business-jet wingshape by a Nash game and an adapted split of variables*. Mécanique & Industries **11**.3-4 209-214, (2010).

[2] B. Abou El Majd. *Optimisation de forme paramétrique: Stratégies hiérarchiques, adaptatives, et concourantes*. Éditions universitaires européennes, (2015).

[3] R. Aboulaich, A. Habbal and N. Moussaid, Split of an optimization variable in game theory, Math. Model. Nat. Phenom(MMNP) **5, No. 7**, 106-111,(2010).

[4] R. Aboulaich, A. Habbal and N. Moussaid, Optimisation multicritère : une approche par partage des variables, **13**, 80-87,(2010).

[5] R. Aboulaich, R. Ellaia, S. El moumen, A. Habbal, N. Moussaid. *A new algorithm for approching Nash equilibrium and Kalai Smoridinsky solution*, submitted to "Applied and Computational Mathematics an International Journal (ACMIJ)".

[6] R. Aboulaich, R. Ellaia, S. El moumen. *The Mean-Variance-CVaR model for portfolio optimization Modeling using a Multi-Objective Approach based on a hybrid method*, Math. Model. Nat. Phenom, **7**, 93-98, (2010).

[7] R. Aboulaich, R. Ellaia, S. El Moumen. *A new hybrid method for solving global optimization problem*, Applied Mathematics and Computation Volume 218, Issue **7**, 3265-3276, (2011).

[8] I. Das, J.E. Dennis. *Normal Boundary Intersection, A New methode for Generating the Pareto Surface in Nonlinear Multicreteria Optimization problems*, SIAM J. Optimization, **3**, 631–657, (1998).

[9] J. A. Désidéri, R. Duvigneau, B. Abou El Majd, and Z. Tang. *Algorithms for efficient shape optimization in aerodynamics and coupled disciplines*. In 42nd AAAF Congress on Applied Aerodynamics, March 2007, Sophia-Antipolis, France.

[10] M. Duran Toksarý. *A heuristic approach to find the global optimum of function*, Journal of Computational and Applied Mathematics **209** 160-166, (2007).

[11] S. Kirpatrick, C.D. Gelatt, and M.P. Vecchi. *Optimization by simulated annealing*, Science,New Series **220**), 671-680, (1983).

[12] H.M. Markowitz. *Multivariate Stochastic Approximation using a Simultaneous Perturbation Gradient Approximation*, **7**, 77–91, (1952).

[13] N. Metropolis, A.W. Rosenbluth, M.N. Rosenbluth, A.H. Teller, and E. Teller. *Equations of state calculations by fast computing machines*, J. Chem. Phys., **21(6)**, 1087-1091,(1953).

[14] B. Raphael , I.F.C. Smith. *A direct stochastic algorithm for global search*, Applied Mathematics and Computation **146**, 729-758,(2003).

[15] C.R. Reeves. *Modern Heuristic Techniques for Combinatorial Problems*, John Wiley and Sons, New York, NY, 1993.

[16] R.T. Rockafeller, S. Uryasev. *Optimization of Conditional Value-at-Risk*, Journal of Risk,**3**, 21–42, (2000).

[17] R.T. Rockafeller, S. Uryasev. *Conditional Value-at-Risk for general loss distributions*, Journal of Banking and Finance, **7**, 1443–1471, (2002).

[18] M. Smorodinsky, E. Kalai. *Other Solution to Nash's Bargaining Problem, Econometrica*, Applied Mathematics and Computation, **3**, 513–518, (1975).

[19] J.C. Spall. *Multivariate Stochastic Approximation using a Simultaneous Perturbation Gradient Approximation*, IEEE transactions on automatic control, **37** 332-341, (1992).

[20] S. Uryasev. *Conditional Value-at-Risk: Optimization Algorithms and Applications*, Financial Engineering News **14**, 1-5, (2000).

[21] Q. Yuan, Z. He, H. Leng. *A hybrid genetic algorithm for a class of global optimization problems with box constraints*, Applied Mathematics and Computation **197** (2008) 924-929.

The elaboration of a manufacturing flow connectivity model, based on Multi Agent System

Lamyae Fahhama[1,*], *Abdellah* Zamma[1], *Khalifa* Mansouri[1], and *Zayer* Elmajid[1]

[1]Laboratory SSDIA, ENSET Mohammedia, Hassan II University of Casablanca

Abstract. Nowadays, manufacturing industries are operating in a highly complex and unforeseeable economic context. Therefore, they tend, by necessity toward short term inter-firms coordination. Thus, found increasingly located at the intersection of corporate networks, to overcome their inability to quickly meet the needs of clients, facing thereby several complications related to the coordination between heterogeneous entities. Managing the flow of a network, made up of different links became a major challenge. The appeals to decision support tools and simulation are wished all the more to correlate the dynamics of the system and the targeted objectives, to assess the configurations effectiveness. Thus, enhance their performance.

The aim of this paper was to establish a model of the industrial flow connectivity; Afterward, we've detailed a network configuration model based on the multi-agents systems, to study the interactions between all the actors and give a more realistic vision onto manufacturing coordination in the supply chain.

1 Introduction

At the beginning of the industrial era, the productivity has been for a long time the principle guiding for a company. Today, with the mutation of the competition and the market globalization, companies became aware that a local effort of improvement of their means of production are not any more enough to reach the level of reactivity and flexibility required to meet the needs of demanding customers. Thus, a tight collaboration and an integration of the various actors of the supply chain turn out, essential to face the exacerbated requests to meet client's satisfaction through the deadline, the cost, the quality of service provided, etc.

Hence, to face these stakes, manufacturing companies, broke more and more of their design activities of production and assembly to the inside of their chain, therefore they're brought to search and benefits from the expertise of potential partners. Thereby, having a multi-site, multi-trades design, so they can better respond to the needs of the customers. Changing from a one-level structure to a complex multi-level configuration, that is manufacturing network.

In this perspective, the configuration of manufacturing networks constitutes a problematic major industrial, more particularly in a context of heterogeneous levels. The piloting of the dynamics of the flow of production and the assessment of the performance of the logistics chain incorporated is crucial. It is concerned more precisely to the configuration of a new link or a company Ein combines with a level below a manufacturing network, also named the connectivity of flows [1]. This organizational coalition can be characterized by a number of attributes describing the operating system and the control one of the manufacturing company. These attributes allow defining firms profile able to integrate a manufacturing network. Therefore, supply chain coordination (SCC) offers means to understand and analyze a SC as a set of dependencies both in physical flows, as well as in informational flows. Our contribution approach is to appraise how efficient the process connectivity in supply chain, made up of heterogeneous, but yet compatible profiles of companies can be. Using communicating agent architecture. The objective of this approach is the anticipation and the evaluation of different scenarios and the detection of the less performing configurations.

Our article is organized as follows: The next section presents a literature review in relation to our work is given; the section 3 introduces our model of flow connectivity in the manufacturing supply chain. In sections 4 we formalize the method of coalitions linking adapted to the distributed coordination by a multi-agent approach. Furthermore, we explain the interactions necessary for its implementation. Finally, we conclude on this work.

2 The modeling and the simulation of flow connectivity in manufacturing networks

2.1 The flow connectivity

In the Supply chain, manufacturer maintains various types of flow. It may be an information flow, relating to supplies or to the design of products, a financial flow, or even a materials flow raw materials and parts of assembly up to finished products. Coordinating the flow between

*Corresponding author: Lamyaefahhama@gmail.com

two companies then returns, to ensure that the information flows reliably and rapidly between them and that the satisfaction of the customer need be realized in time, with the least cost and the minimum levels of inventory [2-3].

The problems of flow management discussed in the literature are focused as well at the global level between several actors in the SC (inter), that within the activities of one of the actors (intra) [4].

Whereas, researchers were interested in the integration and coordination of the actors of the SC, in order to make the flow more fluid, agile and responsive to the dynamics of the market [4-6]. They had proposed solutions for the management of the SC often requiring a modification of the organizational, facilitated by the implementation of the new technologies of information and communication. However, the cumbersome and the difficulty of implementing these approaches, often pose problems. In particular, small and medium-sized enterprises, the fact of the rigidity and the limitation of their financial means, cannot always be involved in projects of major restructuration of their organizations [7].

2.2 The modeling and the simulation of the connectivity

Works the literature are unanimous on the role of modeling to interpret the reality in order to assimilate it, and criticize it; According to Fischwick, 1997 *'The only way to criticize a modeling is to make another by changing the settings'* due to the fact that reconfiguration of the parameters in reality is not always possible. Coupled with the simulation, it allows by the conduction of experiments, based on the predetermined model, to study the possible scenarios and evaluate it to deduce the decisions or of findings.

The approach is adopted in several works in order to reduce the complexity of the management of the supply chain and to assess its performance [8]. The models based on communicating agent for the simulation of SC are in vogue, filling thereby the gaps of analytical models to model the chain with all its complexity [9-10]. It's powered in the other hand, by the simulation making possible to manage the stochastic behavior of the Supply Chain before their implementation, and provide a tool to help decision making, by taking into account the dynamics of the system and of the behavior of the different autonomous actors. [11-12].

Some authors, such as [13] have had recourse to the multi-agent systems to resolve the problems of supply chains. Indeed, [14] that had described a framework based on agents for the SCM. Thus, developing a process of negotiation between functional agents. Other work as [15] designs an Information System for the Multimodal Management on the basis of MA, to that is added the thesis of [16] which employment the DAI for the modeling of Meta heuristic, and [17] which he combines the mathematical optimization and the DAI modeling for the development of its system.

It is obvious that, what has been discussed, in the literature around the connectivity of the process of manufacturing networks, established in a supply chain are, either centered on particular industrial problems, or have a qualitative character and thus not reflecting reality. The taking into account of the special features related to planning modes, control of type of production and response to the demand are rarely discussed. Yet they make this system more efficient (reduce delays and stocks and make consistent links). A better model for process connectivity can be proposed, allowing the assets of the performance in an early phase.

Our contribution fallows the same approach of modeling and simulation, to respond to the problem of the configuration of the manufacturing links, within a manufacturing network.

2.3 The problem description

In related literature modeling approaches have gained importance in supply chain modeling, there exist vast amount of studies aimed at modeling supply chains under multi-agent setting. However, most of the studies focus on conventional supply chain structures and do not provide a real-world application. In this sense, the notion of 'fit' in an existed supply chain management (connectivity) literature has evolved to achieve greater organizational and operational effectiveness. Raison why the supply chains should be modeled as open and dynamic systems by considering complex adaptive relationships between agents.

In parallel multi-agent system modeling is a suitable approach for conveniently modeling the decision making processes by considering complex relationships. With these ideas gathered, we investigate the issue of supply chain coordination that arises in a supply chain and it is necessary to understand and analyze as a set of dependencies both in physical flows, as well as in informational flows.

A simulation model, which is based on the multi-agent system simulation approach, was considered to help and explore the need to manage these dependencies for a company success, ensure higher productivity through fluent and coherent profiles, and reduce the lead-times and inventory levels among the partners.

The purpose is to address a asses to the coalition making possibility to achieve a state of fit between partners configuration settings and performance indicators, while taking into account the need to achieve a higher service level and customer satisfaction (effectiveness) on the one hand, and being cost-efficient on the other hand (efficiency).

3. The proposed model of manufacturing process connectivity

Unlike the problems of cooperation in the supply chains, existing in the literature, interested in the interactions between the actors, the coordination and sharing of information, our model incorporates the preliminary phases of the establishment of the connection, its establishment and the assessment of its relevance to the achievement of the optimization objectives of the overall chain in any coalition making between partners.

Our model show the different relations between the actors of the supply chain, the configuration of the links between the manufacturing enterprises, of the considered manufacturing network and integrates the algorithm of Coalition [18] to meet the needs of the Existant Link, mainly composed of E_{in}. Knowing the mode of response to the clients request and the level of capability of partners.

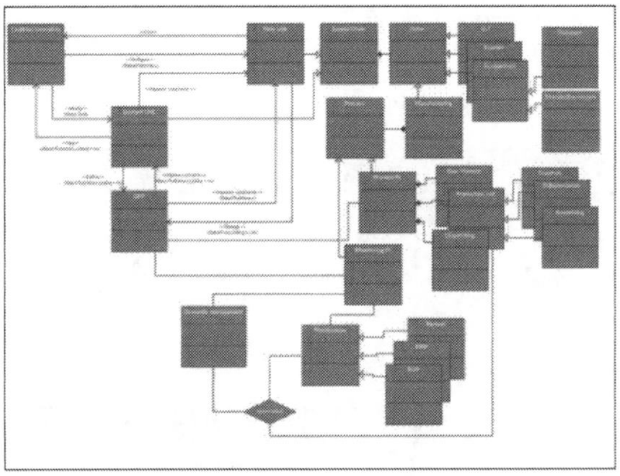

Fig. 1. Class diagram of the connectivity of flow in manufacturing network model.

This modeling will give a better visibility of the problem of connectivity of flows and the configuration of the new links in the same Supply Chain. We briefly present the model proposed of connectivity. We have used the language UML for our modeling and particularly the class diagram (Figure 1).

The connectivity and the need for collaboration, emanates from the position of the process relative to the Order penetration point of command OPP. The characteristics of the products, the type of response to demand, production processes type, planning methods are then identified.
"OPP" is defined from the "Process PH" and the "process of piloting" pre-determined by the "type of response to the client demand" namely MTO, MTS, ATO of the manufacturing company "Manufacturing".

The connectivity of flows under hears the reconfiguration of the chain through the class "formalism of coalition", on which we focus. The class "Existing Link" is the precursor of this configuration, initially composed of an entity of production. In effect, the configuration of the manufacturer's network is considered as a strategic decision requiring a particular intention relative to the performance level wished. The two classes determine and define the vast majority of the constraints to correctly position the types of businesses and thus avoid the design of logistic processes little consistent.

Consequently, the dependence which is established between the Existing Link and the new configured link of the Supply Chain allows à optimized flow stream.

We recall the class "OPP" represents all the constraints related to the configuration of the new link. It is characterized by a Boolean attribute "status", this attribute provides information on the state of the institution or not of a connection. The attribute "status" is a variable which can take two values:
"STATUS = Preliminary", indicating that the coalition referred to the constraints of the E_{in} is in it beginning.
"STATUS = final" indicates that there has configured a new link, responding accurately to the expectations of the existing link.

The class "Supply Chain" is a generalization of the classes "Existing Link" and "New Link" to design.
The existing link is a link that is previously available in the Supply Chain. The configuration must satisfy the constraints of this link while satisfying the entire of the technical requirements imposed by the OPP.

The "New link" is a link formed to answer the needs of the Ein available in the network represented by the "Existant Link" in one hand and on the other hand resulted of the treatment of the "Coalition formalism". It will be determined by taking into account the constraints of the mode of response to the demand of the product by the "Existant Link".

The class "Coalition formalism" represents an adapted algorithm of suitable coalition, for the configuration of new link of the SC. The algorithm used has for objectives to train an adequate coalition in the chain, and make a selection based on preference indexes, calculated in reference to determined criteria (Capability of process..); that so to help choosing the best of the heterogeneous entities in terms of means and modes of production.

The "Existant link" exercising its function in interaction with the other actors in the SC. In fact, it's the one which triggers the process of coalition, based on the constraints of the OPP. This class is characterized by a Boolean attribute 'satisfaction' which provides information on the adequacy and the response to its need, by the chosen configuration. The attribute 'satisfaction' is a variable that reflects the state of this satisfaction:
"Satisfaction = yes" which means that all the constraints of this link are met.

"Satisfaction = no" which means that there is at least a constraint in this link which is not satisfied by the current design.

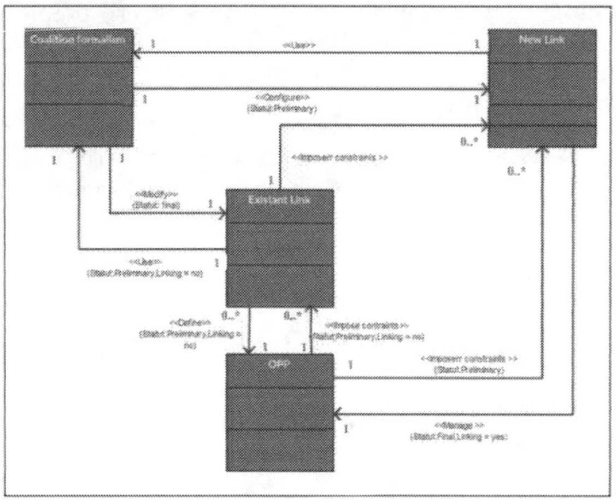

Fig. 2. The class diagram modeling the proposed approach.

The sequence diagram shall present, puts the emphasis on the sequence of messages exchanged between the echelons. It thus, represents the process of coordination in the manufacturing chain between the initiative echelon E_{in} and its potential partners $E_{1,n}$ in order to establish a synchronization of the physical flows in the case of the establishment of a collaboration between the two or more.

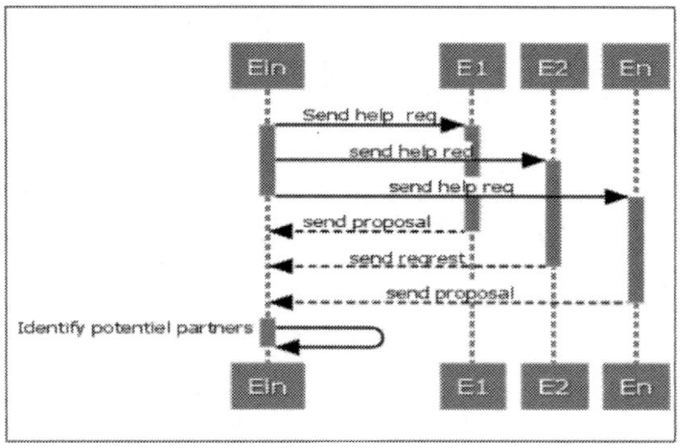

Fig. 3. The sequence diagram for firm collaboration sourcing.

4. The agents based manufacturing network connectivity

4.1 Model main Components

The analogy and the intersection between the characteristics of the MAS with the structure and the dynamic behavior of decentralized, complex and reconfigurable systems [19] as the connectivity of the manufacturers network, has led us to consider a support for decision-making, by proposing an architecture based on the VOWEL approach [20]. This approach models the concepts of the multi-agent system, resulting from the distributed artificial intelligence and the object oriented modeling, by 4 views to know:

The **A**gents view: for the internal architecture of agents, the view **E**nvironment: to describe the environment in which evolve the agents, an **I**nteraction view: the means by which the agents interact and last but not least, the **O**rganization view: the means used to structure the set of entities.

The Topology group of our architecture includes the group configuration of a new link and the Group Supply Chain; we are interested in the first group which includes the whole of the agents participating in the coalition formation .Composed of agents 'coalition', 'Configuration', 'Decision', 'AIC' and Agent 'Existing Link'. The table 1 below contains the agents of the proposed architecture centered on the configuration of the manufacturing network, gives the name, the type and the role of each:

Table 1. Agent identification by type and role.

Agent Name	Type	Role
Existant Link Agent	Réactif	Collect information of echelons, Introduce the connectivity, define the objectives and has a knowledge of the constraints of the OPP,
Configuration Agent	Cognitif	Propose adapted new links based on the information received from the agent **Coalition**. Receives the improvements of the agent **Decision** and transmits the configuration to the agent **AIC**,
Coalition Agent	Cognitif	Analyze the objectives of the existing link and the constraints of the OPP and handles the information received from all other agents to communicate them to the **Configuration** agent,
Decision Agent	Cognitif	Analyze the configuration predefined by the **Configuration** agent and decide of the preservation or not of the connection,
AIC Agent	Réactif	Ensure communication between agents.

To those we add, the group supply chain, through which we model the entities of the SC by agents. To know the agents: Agent 'supplier', Agent 'client', Agent 'Eco-partner', Agent 'manufacturing'.

Each agent has a set of rules and behavior that it implements to succeed this configuration approach of the manufacturing network and this in order to achieve the level of the capability required and an efficient configuration. In what follows, we present our model organization and the interactions of the agents, modeling their dynamic behavior [21].

4.2 The Proposed Agent based Model

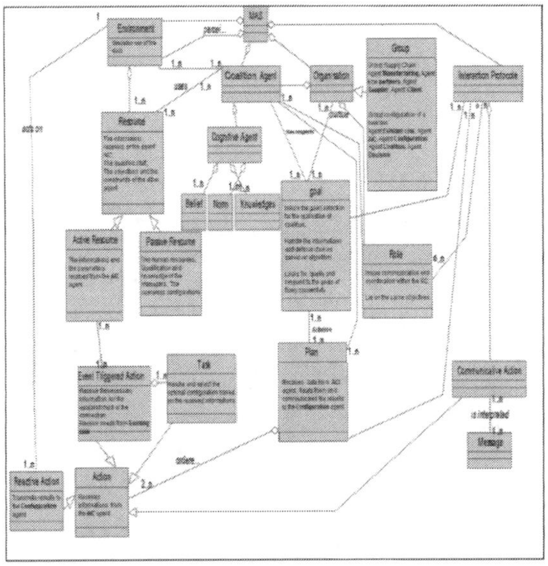

Fig. 4. The global multi-agent model identifies different views of the AEIO approach.

The configuration of the manufacturing network is formalized by the sequence diagram bellow using AUML (Figure 5). The agent 'AIC' receives the information from the agent 'Existant Link' and under hears a communication with the other agents of the Supply Chain. The information comprising the formulation of the specific needs, constraints and objectives including the selection of potential partners, are sent to the agent 'Coalition', where they are treated. The proposition of a selected partner $E_{1,n}$ is sent to the agent 'Configuration' responsible of the configuration of a new link.

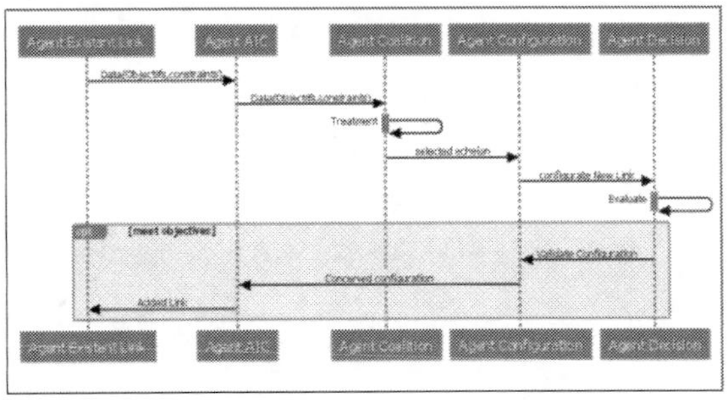

Fig. 5. The sequential diagram of interaction and communication between agents.

The agent 'Configuration', ensures according to the information received from the agent 'Coalition' the reconfiguration of the manufacturing network, the relevance and validity of this new link will be assessed at the level of the agent 'Decision', who can either validate the configuration of the new link, as it may not.
The confirmed configuration will be sent to the agent 'AIC' to be broadcast to the agents of the chain.

This process ended, when the agent 'Decision' validates a configuration responding to constraints (objectives, the adequacy to the processes homogeneity, capability of the SC). The echelon added will be considered as an 'Existing Link'.

5. Summary and outlook

The work presented in this paper is related to the field of integrated engineering, we focused on flow connectivity in the manufacturing network. Managing this heterogeneous form of configurations is mainly based on an information sharing and matching activities between the various links.

In a first step, a systematic literature review has been considered includes different topics related to SCM, its modeling and its simulation in order to explore all types of models, used by several researchers. The aim of this step was to be inspired later, while developing our model presented in the second step.

Thus, we've presented an agent based model of the flow connectivity, the framework able to analyze quality, and effectiveness of configured links and their effects on performance. In addition, we've illustrated the whole interactions between the various agents. The employment of the MAS in linking process addresses many issues where future research could yield interesting results.

Acknowledgement

The authors gratefully acknowledge financial support from The National Center for Scientific and Technical Research (CNRST), Morocco.

References

1. M. Essaid, "Modélisation et simulation de la connectivité des flux logistiques dans les réseaux manufacturiers," p. 207, (2013)

2. N. V. K. Jasti and R. Kodali, *Int. J. Prod. Res.*, no. August, pp. 1–19, (2014)

3. A. Kaddoussi, "Optimisation des flux logistiques: vers une gestion avancée de la situation de crise," *Trait. du Signal*, (2012)

4. D. Anane, S. Pinson, and S. Aknine, "Les Approches Agents Pour La Coordination D'Activités Dans Les Chaînes Logistiques," *Cah. Du Lamsade*, (2009)

5. T. Vadvári and P. Várlaki, "Identification of Supply Chains Based on Input-Output Data," *Period. Polytech. Transp. Eng.*, vol. **43**, no. 3, pp. 162–167, (2015). DOI: 10.3311/PPtr.7931.

6. X. Zhang, "Analysis for Scale-Free Network Characteristics of Logistics Distribution Network," *J. Serv. Sci. Manag.*, vol. **7**, no. 3, pp. 189–195, (2014).DIO: 10.3311/PPtr.7931

7. F. Demoly, X. T. Yan, B. Eynard, S. Gomes and D. Kiritsis, "Integrated product relationships management: a model to enable concurrent product design and assembly sequence planning", *J.Eng.Des.* , Vol. **23**, no 7, pp. 544-561, (2012). DIO:10.1080/09544828.2011.629317

8. Y. Chu, F. You, J. M. Wassick, and A. Agarwal, "Simulation-based optimization framework for multi-echelon inventory systems under uncertainty," *Comput. Chem. Eng.*, vol. **73**, no. 2015, pp. 1–16, (2015). DIO: 10.1016/j.compchemeng.2014.10.008

9. P. Jain , D. Dahiya, "An Architecture of a Multi Agent Enterprise Knowledge Management System Based on Service Oriented Architecture", *IJCSI*, Vol. **9**, Issue 2, No 3,(2012).

10. H. Nfaoui, Y. Ozrout, "SCM, AUML : un modèle d'agents pour la simulation proactive et l'aide à la décision dans le Supply Chain", *Conférence Majestic, MAnifestation des JEunes Chercheurs en Sciences et Technologies de l'Information et de la Communication*. 22-24, Lorient France, (2006)

11. J. Costas, B. Ponte, D. de la Fuente, R. Pino, J. Puche, Applying Goldratt's theory of constraints to reduce the bullwhip effect through agent-based modeling, Exp. Syst. Appl. 42 (2015) 2049–2060.

12. V. Kumar, S. Srinivasan, "A Review of Supply Chain Management using Multi-Agent System", *IJCSI*, Vol. **7**, Issue 5, (2010).

13. W.E. Walsh, M.P. Wellman, "Modeling Supply Chain formation in multiagent systems", *In: Agent Mediated Electronic Commerce (IJCAI Workshop)*, 94–101, (1999).

14. Y. Chen, Y. Peng, T. Finin, Y. Labrou, S. Cost, B. Chu, J. Yao, R. Sun and B Wilhelm, "A negotiation based multi agent system for Supply Chain", *In: Working Notes of the Agents '99 Workshop on Agents for Electronic Commerce and Managing the Internet-Enabled Supply Chain*, Seattle, WA, April (1999).

15. M. A. Kammoun, Conception d'un système d'information pour l'aide au déplacement multimodal : Une approche multi-agents pour la recherche et la composition des itinéraires en ligne. Lille: Ecole Centrale de Lille, (2007).

16. Y. Chu, F. You, J. M. Wassick, and A. Agarwal, "Simulation-based optimization framework for multi-echelon inventory systems under uncertainty," *Comput. Chem. Eng.*, vol. **73**, no. 2015, pp. 1–16, (2015), DIO: 10.1016/j.compchemeng.2014.10.008

17. M. Sghaier, Combinaison des techniques d'optimisation et de l'intelligence artificielle distribuée pour la mise en place d'un système de covoiturage dynamique. Ecole Centrale de Lille, (2011).

18. A. Mortazavi, A. Arshadi Khamseh, and P. Azimi, "Designing of an intelligent self-adaptive model for supply chain ordering management system," *Eng. Appl. Artif. Intell.*, vol. **37**, no. March, pp. 207–220, (2015). DIO:10.1016/j.engappai.2014.09.004

19. W. Wang, W. Fu, H. Zhang, and Y. Wang, "Hybrid modeling and simulation of automotive supply chain network," vol. **6**, no. 9, pp. 1598–1605, (2013). ISSN: 20407459

20. M.Shaiq, R. Shaikh and A.Ahmed, Conflict Resolutions in Supply Chain Collaboration a Perspective of Supply Chain Managers, *Journal of Business Strategies*, Vol. **9**, No.1, pp. 17–30, (2015)

21. A. Seco and C. Vieira, "A multi-agent supply chain simulation analysis through a statistical mixed model," *Procedia Technol.*, vol. **16**, pp. 163–171, (2014). DIO:10.1016/j.protcy.2014.10.079.

Ontology for Risks in Medicines Supply Chain: Case of Public Hospitals in Morocco

Touria Benazzouz*, *Abdelwahed* Echchtabi, and *Abdelkabir* Charkaoui

University Hassan 1st, Laboratory of Mechanics, Industrial Management and Innovation, FST Settat, Morocco

Abstract. For the supply system of medicines in public hospitals in Morocco, ontology is a solution that will improve understanding between actors, promoting information dissemination and their exploitation and promote a new approach to designing information systems. However, implementation of such an approach requires close collaboration between the various partners and requires an intensive exchange of data between them. These technical data must be transmitted, interpreted, managed, reused and stored in a consistent and standardized way. In this work, we developed an ontology using the OWLGRed tool from a UML model whose objective is risk management in the medicines supply chain in Morocco, principally in the public hospitals that supply itself in a centralized way through the supply division which handles the purchase, storage and distribution of pharmaceuticals in these institutions.

1 Introduction

Pharmaceuticals save lives and improve health and play a vital role in many aspects of health care. For this, they should beings available in hospital establishments at all times in sufficient quantities, in appropriate form, with assured quality and adequate information and at an accessible price [1].

Each partner contributes to the availability of pharmaceutical products at the hospitals, needs access to a set of information and knowledge distributed in several information systems (internal or external to its organization) to make effective choices while taking into consideration all the required constraints and risks that may disrupt the smooth functioning of the chain. The partners also need to share decisions and data with other partners that need it. However, each partner in the pharmaceuticals supply chain has its own domain of expertise and each speaks its own language.

With the emergence of ontologies as a new paradigm of information modeling, sharing and exchange of data between the different actors of a supply chain has become possible. Thus, a new integration and interoperability track has become promising. In effect, ontologies allow not only to exchange information between systems without loss of semantics but they also offer the ability to infer new information from existing ones [2].

Our aim in this paper is to use ontologies as integrator element and as efficient support for the identification of risks related to the medicines supply chain at the hospital centers in Morocco (excluding Universities Hospital Centers) to facilitate management and make explicit the common understanding of risks between the different actors of this chain.

To do this, in the 2nd section, we present a literature review of risks associated with the pharmaceuticals supply chain. Then, in the third section, we define some concept related to ontologies. The fourth section will include an analysis of the various risks associated with the medicines supply in public hospitals in Morocco and a description of the ontology developed for manage these risks. Finally, in Section 5 we conclude and present some future works.

2 Risks in pharmaceuticals supply chain in Morocco

The pharmaceuticals supply chain is defined as a set of processes, operations and organizations involved in the discovery, development and manufacturing of pharmaceuticals, which aims to provide medicines with the good quality at the right time and the right place and customers with optimal cot to be consistent with the objectives of the health system [3]. Then, there are risks in a supply chain when unexpected events might disrupt the flow of materials on their journey from initial suppliers through to final customers.

Several studies propose a risk classification and modelization through medicines supply chain ([4] and [5]). The majority of identified risks are similar to those

* Corresponding author: touria.ben.azzouz1@gmail.com

 DOI: 10.1051/matecconf/201710500012

prevailing in industrial supply chains despite the pharmaceutical products particularities. In Morocco, the medicines supply chain (Figure 1) is characterized by a context with large disturbances, a high uncertainty in supplying of medicine and medical devices and particularly by the significant risks which can decrease the coverage of patients needs and generate dangerous consequences for the Health Ministry, the patient, the hospital (reputation, decrease in performance) and for the care team.

In effect, the CESE report in 2013 emphasized that in level the hospital establishments (HE), the needs determination for medicines for the year n+1 is made on the basis of the average consumption of the previous years and not from the real needs and medical prescriptions. This low involvement of prescribers at the time of the needs expression generally produces overestimates and contributes to the accumulation of the medicines stocks and to their expiry.

Also, the CCMPS report (2012) stated that the medicines storage poses problems, with large quantities of expired medicines, for delays in deliveries, rupture of frequent stocks and the insufficient of the logistical means at central and local level. The medicines distribution suffers from a lack of coordination within the distribution network and the absence of an integrated information system for control the activities in medicines supply chain. In addition, the supply division (SD) as responsible for the consolidation of requirements, the tenders' launch, the reception, the storage and distribution of pharmaceutical products, the administrative management as well as the supply monitoring, doesn't have an information system to allow for effective traceability between the various links in the chain which makes the access to the information difficult and generates a set of risk at all levels [6].

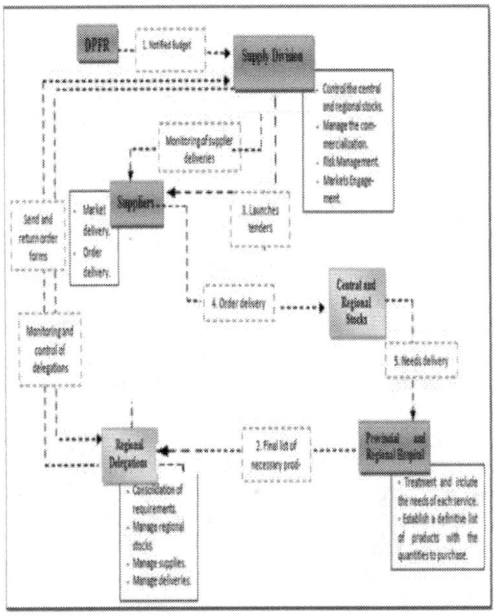

Fig. 1. Medicines supply process for public hospitals in Morocco.

However, several efforts were made to prevent and solve these problems, mainly: the increase in medicines budget by the Health Ministry, the establishment of regional pharmacies, standardization and upgrading of hospital pharmacies and adapting and improving methods of the hospital pharmacy management [7]. And despite all these efforts, the needs coverage of the population in medicines remains insufficient since the patient still complains of the non-availability of these products. So it isn't a question of meeting standards, procedures or standards of good practice, but to identify important concepts in the field, clarifying the relationships between objects to allow standardization of languages between partners.

3 Domain ontology for risks management in the medicines supply chain in Morocco

Domain ontology consists of a shared vocabulary between different partners of the medicines supply chain in Morocco. A first job is to build a corpus, which includes the definition and classification of risk and the relationship between partners. From the vocabulary contained in the corpus, we will extract the constituent primitive concepts of this ontology.

3.1 Definition of Ontology

The ontology is a new modeling paradigm of knowledge borrowed from philosophy and developed as part of the semantic web. They define the concepts, relationships between concepts, choices and constraints that must be respected. In effect, an ontology defines an organized whole (often in the form of taxonomy or semantic network) of usable concepts for formulating knowledge. [8]. Ontologies specify explicitly the conceptual knowledge using a formal or semi-formal language. In the community of Knowledge Engineering, the Ontology term is often associated with a meta-model that describes the contents of a database, its properties, how it can be used and the vocabulary and syntax provided by the representation language.

3.2 Types and construction methods of ontology

There are several types of ontologies and its applications are diverse. We suggest in this context the classification proposed by [9] which proposes four types of ontologies:

- Generic ontologies or upper ontologies : are an independent conceptualization of a problem or a particular area, for example: space, time, objective and event
- Domain ontologies: describe the vocabulary related to an area such as health, industry and education. Also, it can describe the vocabulary associated with a task such as planning, diagnosis or purchase. This type of ontology defines the necessary knowledge to solving a particular type of work.

- Application ontologies: provide concepts based on a specific task and a particular area.
- Representation ontologies: specify conceptualizations that underlie knowledge representation formalisms.

In the literature, there are several works that have proposed the construction methods of these ontologies including the work of ([10-13]. The following table (Table 1) summarizes the definitions of these methods.

Table 1. Construction methods of the ontologies

Methods	Definitions
"Enterprise Ontology" [11]	It is a method which aims: to improve communication between human beings, provide a basis for the specification of user applications and support interoperability.
"TOVE" [14]	A method that necessary to construct of a logical model of knowledge.
METHONTOLOGY [15]	This method is developed by the Artificial Intelligence Laboratory for building ontologies in knowledge level.
La TERMINAE [16]	This method proposes the construction of ontologies from texts.
Unified process for ontology building	This method based on the Unified Process (UP) and UML. UPON is guided by diagrams of use because it aims to build user-oriented ontologies.

3.3 Ontology Development Tools

There are various ontology languages like XML, RDF(S), DAML+OIL and OWL. Many ontology tools have been developed for implementing metadata of ontology using these languages as:

- The Ontolingua server [17] located at Stanford University allows a user or group of users to view existing ontologies and cooperatively build new ontologies.
- WebOnto [18] developed at Knowledge Media Institute at the Open University. It is a web-based tool and mainly graphic to build ontologies cooperatively.
- ProtégéWin [19] was designed for the Department of Informatics Medical University of Stanford, especially for building ontologies.
- ODE [20] ODE (Ontology Design Environment) is a construction support tool for ontology in "knowledge level", which is independent of any formal language.
- OWLGred [21] allows creating, editing and viewing ontology. It offers a comprehensive overview of OWL based on UML.

We aim in this work to use domain ontologies as efficient support risk management in the medicines supply chain in Morocco to facilitate integration, internal and external, of the different partners in this chain. The developed ontology will be built using a UML model. Indeed, we have interest in using the "Unified Process for ontology building" method to identify different risk which may prevent access to medicines and also describe the supply process and partners of this chain. OWLGred tool was chosen to automatically pass our conceptual model expressed as a UML class diagram to a formal semantic ontology represented by OWL.

3.4 Conception of the domain ontology

The domain ontology (DO), as was mentioned before, defines all concepts related to a domain of activities semantically, it concerns here the risk at the level of the medicines supply chain in Morocco. It aims primarily to "assemble and unify the terms related to risk management". Also, it provides a vocabulary of concepts used in the domain with their definition, similar concepts, related concepts, etc. The conception of this ontology was made in the Unified Modeling Language (UML). It is a simple formalism for experts in the domain that facilitates the process of the ontology constitution.

UML profiles used to specialize a UML model. They can be used in all models without modifying their structure. Moreover, it should be noted that such reasoning, although it is possible, isn't exploitable by a machine because it isn't an automatic reasoning since UML profiles don't allow any form of automatic semantic processing. In effect, the domain ontology modeling in UML must then turn into formal ontology. It can then be edited in protected and enriched environment.

Note that several research papers suggest modeling ontologies with UML, including [22]. In [23], the authors develop an approach which aims the visualization of OWL ontologies using the UML graphical notation. To create an ontology with UML (enriched) and then automatically switch to OWL (or vice versa), [24] suggest OWLGred tool. We chose to use the same tool to automatically switch from a conceptual model expressed as a UML class diagram to a formal semantic ontology represented by OWL.

The following figure (Figure 2) shows partners of the medicines supply chain in Morocco (Supply division, suppliers, warehouses, delegations and public hospitals) where each rectangle represents a class, and lines with a triangle represent the generalization relationship between classes.

Identifying needs in hospitals is often left to head nurses or doctors who transfer the patient needs to the service needs (Need Class) and transmit them to the hospital pharmacy (Hospital Pharmacy Class). The Need Class also includes the programmed and non-programmed (urgent) needs for one or more services (Service Class). All these needs are then transferred to the Regional Delegations and Provincial Directorates (Delegation Class) that compile annual commands on pharmaceuticals of one or more hospitals (excluding the CHU : autonomous supply). Then, the commands elaborated send to the SD (SD Class) to begin the process of purchasing from suppliers. The Suppliers class represents laboratories and industrial establishments involved in procurement at warehouses depending on the type of products delivered. At the hospital pharmacy, the products delivered by the central warehouses (Warehouse Class) are then stored in the stock of pharmacy (Stock Class) and distributed to various services according to a schedule adapted for each hospital. The dispensation is made according to the care units type (Service Class), product type (PH Class) and resources available but generally we can distinguish between two main modalities: Global and Nominative. Indeed, the Distribution Class creates a flexibility of choice of the distribution methods (Global and Nominative Classes) of medicines and medical device (Medicines and MD Classes).

The global distribution is to deliver the products to the service (Service Class) in advance on presentation of a weekly order for ordinary commands for a period of a week or an order for pharmaceutical complement which allows the service to order the products which don't available in its stock at any time. The products are delivered just after the deposit of a delivery order signed by the service head. These products are then stored at the service and issued to healthcare personnel (Nurse Class) and (Doctor Class) who administer by a medical prescription. Concerning the nominative distribution, allows the products delivery to a patient (Patient class) and not a service which plays an intermediary role between the pharmacy and the patient, on presentation of a prescription to the nominative delivery of products. Also, the proposed domain ontology allows different partners to identify any risk can be presented in the medicines supply chain into different categories, subcategories and management level (operational, tactical and strategic) (Figure 3).

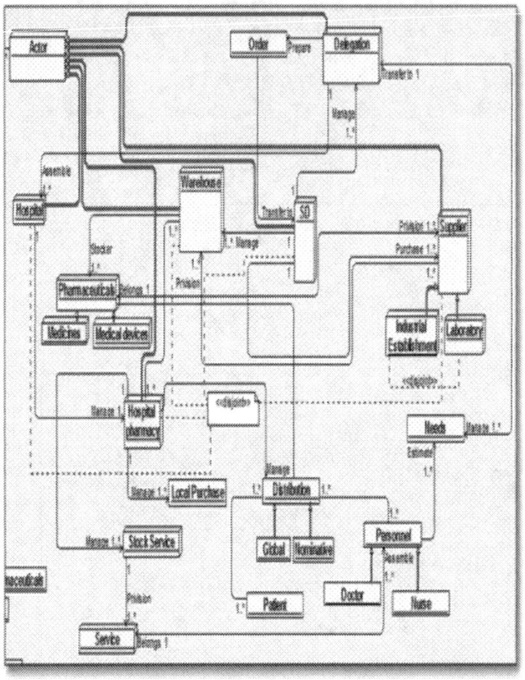

Fig. 2. Demain ontologie : Partners of the medicines supply chain on OWLGred.

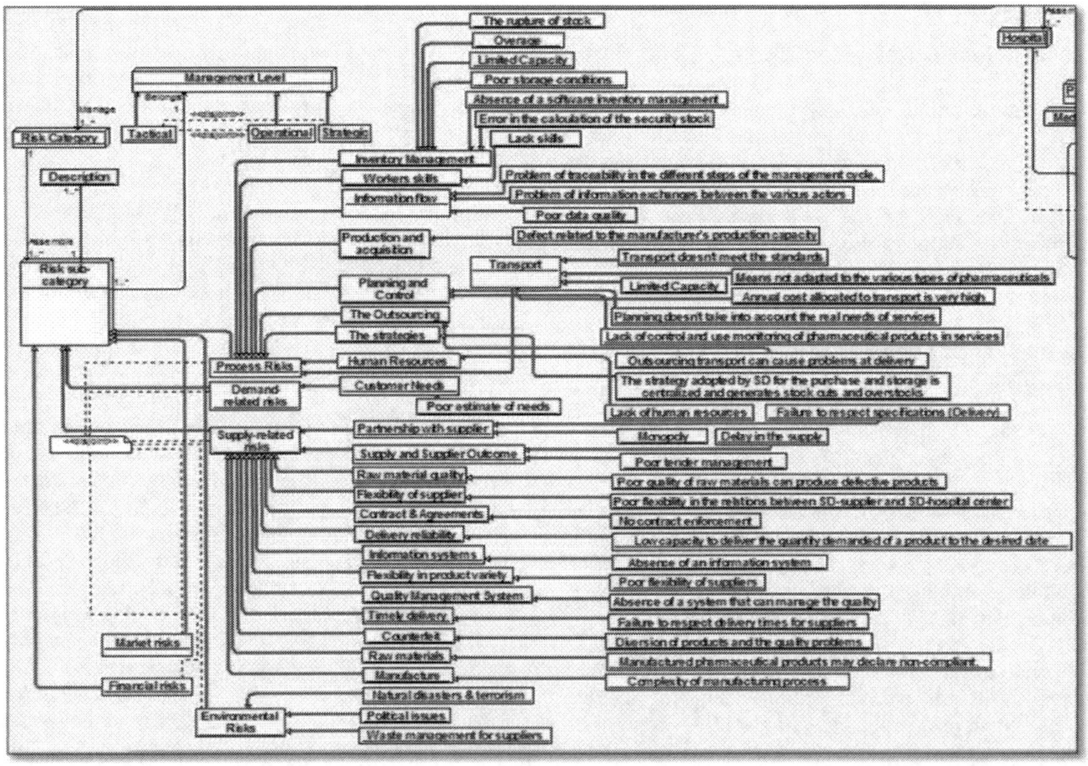

Fig. 3. Demain ontologie: Risk management in medicines supply chain on OWLGred.

4 Conclusion

In this article we proposed domain ontology as a decision support system for the risks management in medicines supply system at the public hospitals of Morocco. The proposed system provides three functions: identifying risks in the medicines supply chain at the level of the public hospitals (except University Hospital Center), semantic description of each risk category and risk management by the various partners in the chain.

The decision support system proposed, will be validated in a larger context with the final users and also the impact of the evolution of ontologies should be studied on this system.

References

1. OMS. Comment élaborer et mettre en œuvre une politique pharmaceutique nationale Deuxième édition. (2002)
2. V. Fortineau. Contribution à une modélisation ontologique des informations tout au long du cycle de vie du produit. Chemical and Process Engineering. Ecole nationale supérieure d'arts et metiers, ENSAM, France, (2013).
3. L. Kaufmann, C. Thiel, A. Becker. Supply chain management in the Mexican pharmaceutical industry. 16th annual North American research/teaching symposium on purchasing and supply chain management (2005).
4. L. Breen. A Preliminary Examination of Risk in the Pharmaceutical Supply Chain (PSC) in the National Health Service (NHS). *J. Serv. Sci. & Management*, **1**, pp. 193-199, (2008).
5. G. Mehralian, A. R. Gatari, M. Morakabati, H. Vatanpour. Developing a Suitable Model for Supplier Selection Based on Supply Chain Risks: An Empirical Study from Iranian Pharmaceutical Companies. *Iranian Journal of Pharmaceutical Research*, **11**, pp. 209-219, (2012).
6. M. Yafout. Informatisation de la chaine d'approvisionnement, état des lieux et perspectives. 1st Meeting of the supply system in pharmaceuticals, Morocco (2015).
7. Health Ministry. Financement public de santé au Maroc entre les contraintes et opportunités (2013).
8. S. ElKadiria, D. Kiritsisa. Ontologies in the context of product lifecycle management: state of the art literature review. *International Journal of Production Research*, **53(18)**, (2015)
9. N. Guarino. Formal Ontology and Information Systems. Proceedings of FOIS'98, Trento, Italy. Amsterdam, IOS Press, pp. 3-15, (1998).
10. T. Gruber. A translational approach to portable ontologies. Knowledge Acquisition. pp. 199-220, (1993).
11. M. Uschold, M. King. Towards a Methodology for Building Ontologies. In Proceedings of IJCAI95 Workshop on Basic Ontological Issues in Knowledge Sharing, Montreal, Canada, pp. 1-13, (1995).
12. N. Guarino. Formal Ontology and Information Systems. Proceedings of FOIS'98, Trento, Italy. Amsterdam, IOS Press, pp. 3-15, (1998).
13. T. Grubic, S. Fan. Supply chain ontology: Review, analysis and synthesis. *Computers in Industry*, **61(8)**, pp. 776-786, (2010).
14. M. Gruninger, M.S. Fox. Methodology for the Design and Evaluation of Ontologies. IJCAI'95, Workshop on Basic Ontological Issues in Knowledge Sharing, (1995)
15. M. Fernández-López, A. Gómez-Pérez, A. Pazos. Building a chemical ontology using methontology and the ontology design environment. *IEEE Intelligent Systems & Their Applications*, **4(1)**, pp. 37-46, (1999).
16. B. Biébow, S. Szulman. TERMINAE: A linguistic-based tool for the building of domain ontology. In Proceedings of the 11th European Workshop on Knowledge Acquisition, Modeling and Management: EKAW'99, Dagstuhl (Germany), (1999).
17. A. Farquhar, R. Fikes, J. Rice. The Ontolingua Server: a Tool for Collaborative Ontology Construction. Proceedings of the Tenth Knowledge Acquisition for Knowledge-Based Systems Workshop. Banff, Canada, (1996).
18. J. Domingue. Tadzebao and WebOnto: Discussing, Browsing, and Editing Ontologies on the Web. In B. Gaines and M. Musen (editors), Proceedings of the 11th Knowledge Acquisition for Knowledge-Based Systems Workshop, April 18th-23th, Banff, Canada, (1998).
19. H. Eriksson, R. Fergerson, Y. Shahar, M. A. Musen. Automatic Generation of Ontology Editors. Proceedings of the 12th Banff Knowledge Acquisition Workshop, Banff, Alberta, Canada, October 16-22, (1999).
20. A. Farquhar, R. Fikes, J. Rice. The Ontolingua Server: Tool for Collaborative Ontology Construction. *IJHCS*, **46(6)**, pp. 707-728, (1997).
21. J. Barzdins, G. Barzdins, K. Cerans, R. Liepins, A. Sprogis. OWLGrEd: a UML style graphical notation and editor for OWL 2. In 7th International OWLED Workshop, (2010).
22. IBM. Ontology definition metamodel, (2006).
23. S. Brockmans, R. M. Colomb, P. Haase, E. F. Kendall, E.K. Wallace, G. Tong Xie. A Model Driven Approach for Building OWL DL and OWL Full Ontologies. The Semantic Web - ISWC 2006, **4273**, pp. 187-200, (2006).
24. J. Barzdins, G. Barzdins, K. Cerans, R. Liepins, A. Sprogis. OWLGrEd: a UML style graphical notation and editor for OWL 2. In 7th International OWLED Workshop, (2010).

MATEC Web of Conferences **105**, 00013 (2017) DOI: 10.1051/matecconf/201710500013
IWTSCE'16

Airline Schedule Disruption Management The impact of flight delays on connection loss

Hicham Rahil[1*], *Badr Abou El Majd*[1] *and Mohammed Bouchoum*[1]

[1]LIMSAD, Hassan II University, Faculty of Science, Casablanca

Abstract. Air travel demand is important and many travellers choose to drive to larger airports instead of flying from a small airport for many reasons, especially availability of non-stop flights. Another reason is perceived reliability of service. Consultants have pointed to a large number of delays and cancellations as reasons for low passenger. However, the effect of these flight delays on actual travel times is less clear. Because connections are usually necessary when traveling from small airports, departure delays may lead to missed connections. In the case of a cancellation, need to wait several hours (often overnight) for the next flight due to the small number of daily departures. This paper evaluate the impact of delays and cancellations on the profit earned through the seats captured on new opened routes. This aspect of decision-making comes in the form of multi-objective problem by testing the impact of a new opened route in terms of flight delays costs, financial gain and the quality of the service provided to a target customer. The NSGA-II algorithm is adopted to generate a front of Pareto-optimal compound of a number of optimal departure times to the new destination while ensuring the best fill rate, and a minimum flight delays. The experiences are based on the flights of the Royal Air Maroc Company on the Casablanca hub.

1 Introduction

In Air Transport field, the most important factor sought by the potential passenger is the shortness of the trip. Passengers therefore preferred the direct flights between the airport of origin and destination. When the passenger is obliged to carry out a correspondence, airlines have an incentive to attract these passengers to increase the profitability of their lines by providing a continuous service from airport of departure, passing via the connecting airport until reaching the final destination. Royal Air Morocco (RAM), the Morocco's flag carrier, hopes to take place among the major airlines pulling profits from the position of the Morocco being the Africa portal on Europe and well obviously at the air hub put at his disposal in Casablanca. In addition to its flights to more than 80 destinations, Royal Air Morocco now offers more flights around the world by many airline partners. An alliance would benefit the position of RAM to the Morocco, but also central Africa / Western, the largest international market of RAM after Europe. In this work, we are interested in the problem of insertion of the new departure and arrival times to the existing program of an airline, and we opted for multi-objective optimization to solve this problem, the multi objective optimization meets the need to satisfy conflicting requirements. Therefore, to travel from Casablanca to Agadir, we will seek to minimize the cost; the time spent and maximize comfort. The plane will be cheap, fast, and polluting while the car will be economic, long and low in pollutants. Finally, the train will be a compromise solution. As there are no better in every way than another solution, a different compromise according to persons must be chosen. The choice is subjective, and it is essential to provide the set of choices so do not exclude a possibility. Multi objective optimization is primarily a tool for decision support, and it is a person who will make the final decision.

2 Outline of the paper

This article will focus on the application of optimization algorithms evolutionary (EA) using air routes scheduling problems and provide a brief overview on how airlines use these techniques to develop and assess the decisions of schedule amendments required at the last minute. Section 2 of this document provides a mathematical formulation of the problem of adding new lines in an existing program, under specific constraints, cost and profitability associated with a given calendar change. Section 3 explains the optimization process to fix the problem. Section 4 will

* Corresponding author: hichamrahil@gmail.com

provide experimental results that explain more and detailed model and the optimization process, discuss possible extensions of the other sketches and model of research. Section 5 concludes the paper and specifies certain points of view. At the end of the paper, we provide a complete list of references highlighted in the document.

3 Methods

3.1 Problem Formulation

A wide variety of problems in industry, and many other fields, involve the simultaneous optimization of several objectives. In many cases, the objectives are defined in incomparable units, and they present some degree of conflict among them (i.e., one objective cannot be improved without deterioration of at least another objective). These problems are called Multiobjective Optimization Problems (MOPs). Let us consider our problem of new route insertion in an existing flight program, where the air transport company is interested in minimizing the total duration of the generated delays over the hub to improve customer service. On the other hand, the company also wants to maximize the number captured seats in order to improve the market benefits. Clearly, these objectives are in conflict since capturing more seats increases benefits, but increases generated delays. In addition, the objectives of this problem are expressed in different measurement units. The method most commonly adopted in Multiobjective optimization to compare solutions is the one called Pareto dominance relation [1], which, instead of a single optimal solution, leads to a set of alternatives with different trade-offs among the objectives. These solutions are called Pareto optimal solutions or non-dominated solutions. Although there are multiple Pareto optimal solutions, in practice, only one solution has to be selected for implementation. Therefore, in the Multiobjective optimization process we can distinguish two tasks, namely: i) find a set of Pareto optimal solutions, and ii) choose the most preferred solution out of this set. Since Pareto optimal solutions are mathematically equivalent, the latter task requires a Decision Maker (DM) who can provide subjective preference information to choose the best solution in a particular instance of the Multiobjective optimization problem. In this paper, the company wants to open a new destination, and the purpose is of course the conquest of new markets. Indeed, after the opening of a new air route on departure from its air hub, the company will seek to connect this new destination with other airports served by the same company. A study of the market is developed beforehand to set among airports served by the company, those that present important potential of passengers wishing to go to the new destination via a connecting flight through the hub of the company. Let us note by x the departure time toward the new destination. Actually, x is the decision variable of the problem, which will affect the two criteria considered in the problem. We will also note by "t" the total duration of the flight to the new destination, and of course the value $x+t$ is the scheduled return hour from the new destination.

We will limit our study to the 20 first airports ranked by origin-and-destination passenger volume and the busiest one is airport k where $k = 1$. However, decision-makers may extend targets considered according to demand and importance of the proposed destination. Let A_k (resp. D_k) the set of the weekly arrivals time (resp. departures time) from an airport $k \in K$ ($K = \{1,2,..,20\}$) to the hub, where K is the set of twenty airports served weekly by RAM company and ranked by origin-and-destination passenger volume. a_{ki} the ith arrival time (resp. d_{kj} the jth departure time) with $i,j \in \{1,2,..,N_k\}$. a_{ki} and d_{kj} are ranked in ascending order), and N_k is the weekly frequency of routes between hub and given airport $k \in K$, we take the minute as unit of time. We note that, arrivals and departures time during the week are included in the interval $[0, 10080]$.

$$\alpha_{ki} = \begin{cases} 1 \ \textit{Outbound connection is realized for } a_{ki} \\ 0 \ \textit{Otherwise} \end{cases}$$

$$\delta_{kj} = \begin{cases} 1 \ \textit{Intbound connection is realized for } d_{kj} \\ 0 \ \textit{Otherwise} \end{cases}$$

w_k are the weight vectors given by:

$$w_k = \frac{P_k}{\sum_{i=1}^{20} P_i} \quad \text{and} \quad \sum_{k=1}^{20} w_k = 1$$

P_k is the origin-and-destination passenger volume recorded in a year between a given airport k served by the RAM company and the new destination, $\sum_{i=1}^{20} P_i$ is the sum of the passenger volume of the twenty busiest air routes served by RAM, ranked by origin-and-destination passenger volume. This parameter is introduced to encourage the connection of airports with important passenger volume. $x \in [x + t + t_{min}, 10080- t_{max}]$, where t_{min} is the minimum connecting time, and t_{max} is the maximum allowed connecting time.

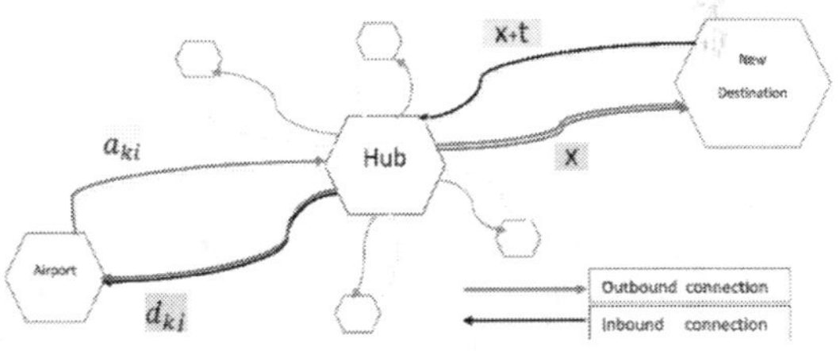

Fig. 1. The Inbound step, pass through the hub to join the new destination, and return for the outbound step.

Figure 1 illustrates the Outbound and Inbound connections, where the flight Inbound step (red color), pass through the hub to join the new destination, and return back again via the hub for the Outbound step (black color). During the optimization process, each random value x generates two intervals: $[x - t_{max}, x - t_{min}]$ and $[x + t + t_{min}, x + t + t_{max}]$. An Outbound connection is realized when a given value a_{ki} comes inside Arrival time window $[x -$

t_{max}, $x - t_{min}]$,($\alpha_{ki} = 1$) . In addition, the Inbound one is realized when a given value d_{kj} is cached inside departure time window $[x + t + t_{min}, x + t + t_{max}]$, ($\delta_{kj} = 1$).

$ST a_{ki}$: is the actual time of occurrence of the estimated arrival a_{ki} ;

$ST d_{kj}$: is the actual time of occurrence of the estimated departure d_{kj} ;

C_{ki} : is the unit time delay cost of arrival aki ;

C_{kj} : is the unit time delay cost of departure d_{kj} ;

We note $X_{ij}^{k} = \alpha_{ki} \times \delta_{kj}$ where

$$X_{ij}^{k} = \begin{cases} 1 & Full\ connection\ is\ realized\ for\ a_{ki}\ and\ d_{kj} \\ & 0\ Otherwise \end{cases}$$

3.2 Multiobjective functions

In order to find the best departure time *x*, we maximize the objective function f_1 and minimise f_2 :

$$f_1(x) = \sum_{kij} w_k \times X_{ij}^{k} \quad (1)$$

$$f_2(x) = \sum_{kij} [\, C_{ki}\ X_{ij}^{k}\ (ST a_{ki} - a_{ki} - \Delta\tau) + C_{kj}\ X_{ij}^{k}\ (ST d_{kj} - d_{kj} - \Delta\tau)\,] \quad (2)$$

A good choice of x, will enhance the value of the Objective function f_1 will be maximized with guaranteeing the connection in both directions, Outbound and Inbound through the factors α_{ki} multiplied by δ_{kj}, and by promoting the best passenger volume through w_k. f_1 is expressed through scalarization method, the weighted sum (WS) method. The WS method is the commonly used scalarization method because of its simplicity, ease of use, and direct translation of weight into the relative importance of the objectives [2]. Its drawbacks are also well known and discussed in literature [3]. These include the followings:

- ✓ It misses solution points on the non-convex part of the Pareto surface;
- ✓ Its diversity cannot be controlled, therefore even distribution of weights does not translate to uniform distribution of the solution points;
- ✓ The distribution of solution points is highly dependent on the relative scaling of the objective.

For efficient results, the objective function f_2 calculates only the flight Delay Losses of connected flights. Thus the found results will help to quantify the impact of cumulated delays caused by these connections, and then will give us an idea about the resulting costs if we opt for this strategy. Finally, the minimization of f_2 will involve the reduction of the generated delays and will make subsequently the correspondence more comfortable for passengers, and therefore, will automatically reduce the costs related to passenger accommodation and all other related charges.

f_2 is the Objective Function of Flight Delay Losses. In most cases, either the arrival or departure flights, the actual arrival time is almost impossible to be consistent with the estimated time, so we make the following provision. The following two situations are not treated as delay: (1) the actual time is earlier or later than the estimated time within **Δτ**: (2) all the flights, which are in advance of the estimated time landing or taking off.

4 Numerical Implementation

4.1 NSGA-II Algorithm

The evolutionary optimization algorithm used to validate numerically the theoretical framework described above is called the Non-dominated Sorting Genetic Algorithm (NSGA-II) [4, 5]. NSGA-II uses a fast non-dominated sorting approach, with complexity $O(MN^2)$, an elitism approach, and uses the scaled crowding distance as a secondary diversity criterion, that requires no additional parameter, and figure 2 describes the NSGA-II procedure. All design variables here are continuous, and the variation operators have been chosen accordingly, as in described by Deb [6] . More precisely, all experiments use the simulated binary crossover operator (SBX) and the polynomial mutation, which can handle directly the box constraints. The other parameters of the algorithm are the population size, which is equal to the archive size, the probability of crossover, the probability of mutation, the dispersion coefficients, and the number of generations before termination. NSGA-II has been chosen here for its long record of successes, demonstrating its robustness to find a good approximation of the Pareto front. Further work will investigate whether other MOEAs outperform NSGA-II on this problem. A good candidate is SMPSO [7] that has excellent performances on continuous problems.

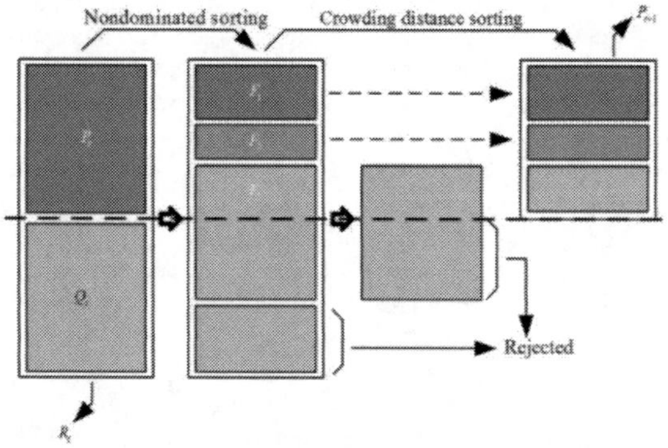

Fig. 2. NSGA-II procedure.

4.2 Results and discussion

Following the air carrier demand, we focus on the test of 20 airports; obtained results are given by figure 3 showing Pareto's front solutions with the corresponding coordinates. We obtain six optimal solutions, offering the possibility to connect 20-targeted airports to the new opening, maintaining an interesting fill rate and minimizing the stopover time over the hub.

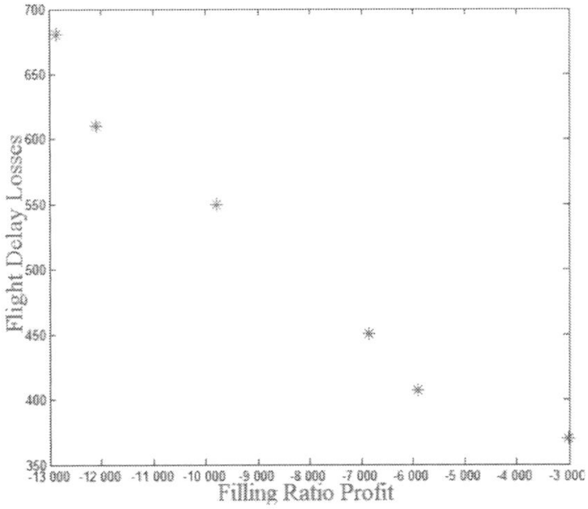

Fig. 3. Pareto optimals of twenty airports case.

Following the air carrier demand, we focus on the test of 20 airports; obtained results are given by figure 2 gives Pareto's front solutions with the corresponding coordinates. We obtain six optimal solutions, offering the possibility to connect 20-targeted airports to the new opening, maintaining an interesting fill rate and minimizing the effect of flight delays. Table 1 gives details of flight rotation corresponding to each solution from Pareto front. We notice that Airports ranked in following orders 15, 16, 17, and 18 are not connected by the obtained solutions. S_4 and S_5 are very interesting solutions, connecting 12 (respect 11) of the 20 tested airports. Solution S_4 provides 10h35min maximum stopover time on the hub of Casablanca in outbound step, and less than one hour (55 minutes) as minimal stopover time. Departure time is Saturday at 06:30 UTC from Casablanca, the return is the same day at 23:00, and the inbound step is 1 hour 15 minutes stopover time in Casablanca and 08h20min maximum before continuing correspondences back to connected airports. Obtained Departures from Casablanca fits perfectly the fleet programming strategy adopted by Royal Air Maroc Company. Indeed, the company sends daily almost the two-thirds of its fleet on Africa within half an hour before midnight to return in the morning on the Casablanca hub before continuing to Europe, America and golf countries. Moreover, knowing that the best potentials of this example are African countries friends and allies of Morocco, the obtained arrival from the new opening around 23:00 (UTC) is perfectly convenient to the fleet departures to Africa starting around 23:30 (UTC). Similarly, a departure time toward the new opening at 06:30 (UTC) is a perfect knowing that the vast majority of those arrivals lands in Casablanca before 06:00 (UTC).

Table 1. $(f_1;\ f_2)$ Solutions with correspondent Rotations

Optimal Solutions	Flight Rotation (UTC)	
	Departure Time	Arrival Time
S_1	Monday - 05:00	Monday - 21:30
S_2	Thursday - 6:40	Friday - 09:10
S_3	Thursday - 00:50	Thursday - 06:40
S_4	Saturday - 06:30	Saturday - 23:00
S_5	Tuesday - 06:25	Tuesday - 22:55
S_6	Tuesday - 12:05	Wednesday - 4:35

5 Conclusions and Future works

Airline managers and decision makers in general, are probably interested to examine the results of this paper, allowing them a dynamic planning of their fleets without need to review the existing flight schedule. Indeed, when we take the decision to open a new destination, the reason may be political or purely commercial; we will be able with this model to insert this new program without losing scheduled connections and without causing major delays during the correspondences. We are all aware that the delay is a boring parameter and causes additional and significant costs for the airlines. Then this solution will serve as a tool to support the decision for an air carrier to amend its program and even think about increasing the number of frequencies on a specific route due to an unforeseen event without constraint to bring changes to the initial program. The Pareto front has been achieved using the algorithm of NSGA-II and the numerical results are based on the real case of the program of the Moroccan airline Royal Air Morocco (RAM) on the hub of Casablanca. Finally, we are all convinced that the aerospace industry is one of the most affected sectors by the hourly disruptions and delays due to several parameters. Airlines put and after developing, a program will face delays that spread quickly to connected flights, which induces considerably important and unexpected costs for airlines [8]. The primary goal now is to seek an effective way to minimize the impact of disruptions, reducing delays and the number of flights affected. In addition and for the perspective of this paper [9, 10], we will include the criterion of disruptions to introduce the aspect of robustness.

References

1. Pareto, Vilfredo. *Cours d'économie politique*. Vol. **1**. Librairie Droz, (1964).
2. Marler, R. Timothy, and Jasbir S. Arora. *"The weighted sum method for multi-objective ptimization: new insights."* Structural and multidisciplinary optimization **41**.6 (2010): 853-862.

3. Das, Indraneel, and John E. Dennis. *"A closer look at drawbacks of minimizing weighted sums of objectives for Pareto set generation in multicriteria optimization problems."* Structural optimization **14**.1 (1997): 63-69.
4. Kalyanmoy Deb, *Multi-Objective Optimization using Evolutionary Algorithms*, John Wiley and Sons, 5 juil. (2001) - 497 pages.
5. Zitzler, Eckart. *"Evolutionary algorithms for multiobjective optimization: Methods and applications."* (1999).
6. Deb, Kalyanmoy, et al. *"A fast and elitist multiobjective genetic algorithm: NSGA-II."* IEEE transactions on evolutionary computation **6**.2 (2002): 182-197.
7. Kennedy, James. *"Particle swarm optimization."* Encyclopedia of machine learning. Springer US, (2011). 760-766.
8. Guimarans, Daniel, Pol Arias, and Miguel Mujica Mota. *"Large Neighbourhood Search and Simulation for Disruption Management in the Airline Industry."* Applied Simulation and Optimization. Springer International Publishing, (2015). 169-201.
9. Abou El Majd Badr, J.-A. Desideri, and A. Habbal, *Aerodynamic and structural optimization of a business-jet wingshape by a Nash game and an adapted split of variables*, Mec. Ind., vol. **11**, no. 34, pp. 209-214.
10. Abou El Majd Badr and S. Kachroudi. *A deterministic approach using game theory for multi-objective robust optimization*, In the proceeding of the 3rd International Conference on Metaheuristics and Nature Inspired Computing, META'10, Djerba Island in Tunisia.

Evaluating the HR function as being a factor of supply chain performance: Case of industrial companies in Morocco

Asmaa Elmortada [1,*], *Ahmed* Mokhlis [1], *Said* Elfezazi [1] and *Chams Eddoha* Mokhlis[2]

[1] LAPSSII laboratory, Cadi Ayyad University, Higher School of Technology of Safi, Safi, Morocco
[2] Laboratory of Scientific Engineering of Organizations, National School of Commerce and Management, Casablanca, Morocco

Abstract. In an economic environment increasingly complex and competitive, firms are now aware, more than ever, of the importance of controlling the human resources function as a primary factor to the success and performance of their supply chains. Through studying the Moroccan industrial companies, the concern of this article is to highlight the role of human resource function in the supply chain performance, in the light of the company culture and under the effect of a HRM practices influenced by the competitive environment, and the resulting mutations.

1 Introduction

As part of the supporting program of the strategic development planning, and in order to develop scenarios for Morocco of 2030, the High Planning Commission conducted a study on "the growth sources of economy in Morocco". This study revealed that the human capital is a key factor for growth by affirming: ***"Human capital is an important ingredient of long-term growth; it [...] can counteract the decreasing returns to scale of other accumulative production factors - such as physical capital -. and sustain long-term growth. Apart from its direct role as a factor of production, human capital can serve as a complement to other resources and factors such as physical capital and natural resources"*** **[1]**.

This context highlights the required reactivity of organization forms particularly those of productive nature by mobilizing and combining diverse skills. This organizational work depends on the ability of the company to attract and mobilize competent and motivated human capital. Good human resource management still then an undeniable source of value creation that places man and his creative potential in the center of all managerial and organizational concerns of a company.

2 Relationship between supply chain and HRM strategies

Having been defined in several ways over the last 40 years, the modern logistics is understood today as a function of planning, implementation and mastery of flows and stocks in the company. It relies on the implementation of information and communication systems increasingly sophisticated and takes place in the supply chain defined by Mentzer as *"a set of three or more entities (companies or individuals) through which flow upstream and downstream products of services, information and finance, from a supplier to a customer"* [2]. In terms of purpose, NFX 50-600 standard specifies that the purpose of the logistic function is to *"meet the needs expressed or latent, internal or external, to better economic conditions for a specific level of service"* [3]. This definition makes that the notion supply chain can accommodate different realities as proved by Chen & al [4]. Management practices flow may also differ from one chain to another, according to the way that the relationship in the supply chain is oriented: Relational Or transactional. Relational relationship is characterized by content (beginning, signals, end) of the communication between the partners in the chain and interaction's mode (speed and frequency), allowing enterprising sides to get goals which are unattainable individually. In ttransactional relationship, the terms of exchange between buyers and suppliers must be specific (low flexibility of suppliers). The benefits and costs of each member of the supply chain are clearly defined (low ease of provider); little information flow between the parties (poor communication); members of the supply chain have little control over each other (low control); transactions are short-lived and have little chance to train new (sustainability little intended). According to the vocabulary of sale, each member is satisfied with a result of "win-lose" provided that they are winning.

Therefore, despite of the challenges that the company may face [5], the guidelines of the supply chain must be adapted to the culture of the company that comes into

*Corresponding author: asmaa.elmortada@gmail.com

mutual influence with HR policies. Indeed, decisions on human resource management have a significant impact because if staff adapts to the culture of the company, it helps to highlight social knowledge, as it helps in promoting cohesion between the objectives of the employee and those of the company. The fact that "Sharing a social conscience helps employees to make the right decisions in the face of changing circumstances" has been proven years ago [6].

In comparison to relations accommodated within the supply chain, human policies and the resulting culture can also be defined according to the Continuum transaction-relation. In other words, we distinguish the relationship-based HRM strategy which emphasizes the reciprocal loyalty between employees and the company and leads to a mutually beneficial relationship in the long run. And the transaction-based human resources strategy according to which employees are interchangeable and replaceable so that much less effort is provided to maintain their loyalty.

By combining the two logistic strategies with the two human resource strategies that a company can adopt, two of the four possible combinations are found to be compatible with both logistic and HRM objectives ("relational logistic-relational HRM" or "transactional logistic- Transactional HRM") while the other two combinations represent a strategic conflict:

2.1 Combined relational strategies:

Through adopting this type of strategy, the company's objective is to establish long-term cooperation with the majority of partners (customers, suppliers and distributors) and its own employees. A relationship-oriented firm strives to ensure coherence and complementarity between vendor and distributor' objectives in terms of human and marketing relationships, thus treating customers, suppliers and distributors alike leads to achieve a high level of cultural coherence that fosters mutual trust between the various enterprising parts (company, employee and different members of the supply chain).

In case of combined relational strategy, employees are seen as investments where the importance of shared responsibility and the "employee" factor dominate the human resource management (lifetime jobs, training, high wages,etc). Choi and Joung conducted a study based on the concept of combined HRM and supply chain management strategies [7].

2.2 Combined transactional strategies:

A company can also adopt a transactional approach regarding some or all of its supply chain partners as well as regarding its own employees. In this configuration, the loyalty of the employee towards the organization and the members of the supply chain is less considered. In this context, fidelity between a company and its suppliers or distributors is less solicited too.

In this scenario, assessing employees is based on the workload to quality dependency criterion. This results in the fact that employees are strictly defined and managed to achieve daily tasks with minimum qualifications. This leads to situations where employees are replaceable and low paid as it can also lead to a further reduction in the efficiency of the services concerned after each personnel change.

2.3 Mixed strategy: long-term relationship with the supply chain members and transactional relation with employees

In this situation, the strategy of the supply chain is based on relationships while the human resources policy is transactional. Members of the supply chain may then receive transaction-level service. This is due to the fact the transactional HRM strategy prevents staff from providing a comprehensive and long-term service, especially since employees are not rewarded for the time dedicated to Maintaining long-term relationships because the reward system is linked to quantitative and transactional objectives. On the other hand, the HR strategy is based on simple and routine jobs that require a low-skilled workforce. Hence, employees are less able to provide the level of service required in a real relational environment.

2.4 Mixed strategy: short-term transaction with the supply chain members and strategy based on relationships with employees

In this case, the company may consider that the members of the supply chain are interchangeable, but it is attempts to get a long-term commitment from the employees. Despite the fact that maintaining a first-class workforce (skilled employees, continuing training and receiving high wages, etc.) for transactional works is one of the less economical aspects of this approach, but it lets the firm able to perform tasks that a member or a transaction chain is not capable of.

3 Objective of the study

The aim of this study is to highlight the mutual influence between supply chain performance and HRM practices through studying the example of industrial Moroccan companies.

4 Methodology

In this article we used a mixed method. The first part of our findings is based on the content analysis of some face to face interviews that we conducted among some HR managers.

We will also present an empirical part which is based on a quantitative research that we conducted among 109 managers based on the administration of a questionnaire. This search method requires the use of mathematical and statistical analysis tools to describe and explain the results through concepts operationalized in the form of measurable variables.

5 Participants

109 persons participated in this study. They belong to different sorts of companies: When 20.18% of these firms are subsidiaries of multinational companies, 22.94% (25) of them are very large companies (over than 1000 employees), 21.10% (23) are large businesses (500-999 employees), 18.35% (20) are medium-sized (50-499 employees), 11.93% (13) are small enterprises (10-49 employees), and 5.50% (6) are very small businesses (1-10 employees). These companies operate in various sectors including the food / fishing area (24.8%), building construction (19.3%), automotive industry (11%), and textile industry (11%), while 8.3% are mining or producing energy (9.2%). Among these enterprises, 14.7% have other activities such manufacturing health products, cosmetics, handicrafts ..., or product packaging. Eighty (73.4%) companies export their products. We made sure that all types of businesses, as well as different business sectors are represented.

6 Sample and procedure

To conduct this study, we searched first for contacts. As part of this, we visited the Chamber of Commerce and Industry which allowed us to have a database containing details of several leaders. We also visited some centers of national unions, in addition to our personal contacts.

On one hand, in the qualitative study we conducted structured interviews while using closed and semi-closed questions. This type of interviews ensures the levels of reliability and high precision, as it allows a better structuring of the debate, especially in the presence of several articles to cover.

On the other hand, during the empirical study we conceived and sent two versions of the questionnaire to participants' email addresses: a Web-based Survey, and a Word file questionnaire. This is done in order to allow the respondent to choose the form that is more comfortable for him or her to use. A cover letter explaining the purpose of the study, and ensuring the confidentiality of the information they can deliver, was also sent to the participants. At the same time, the participant himself can ensure the confidentiality of his answers by choosing the web-based survey which is anonymous.

Generally, we selected respondents using a reasonable choice sampling. This sampling method makes resort to the researcher's judgment to select sample items (Thiétart, 2003). As we are interested in this study to top managers, the participant status was one of the main selection criteria, in addition to the representation of all business sorts and sectors related to the industrial activity, as well as the various geographical regions of Morocco.

7 Measures

Our evaluation of HRF is based on the model of Dave Ulrich (1997) that assigns four main roles to the HRF: administrative expert, employees' champion, the change agent, and the strategic partner. This typology has been used by several researchers, including French, such as Gilbert and Charpentier (2004), or Peretti (1998).

Therefore we proposed subvariables to evaluate the four roles of HRF. For example the variable Documentation (α = 0.862) is intended to assess the 'administrative' expert role (α=0.844); employee incentive evaluates the role of employees' champion, communication assesses the role of change agent (0.853), while the participation of HR managers in strategic decision-making is intended to evaluate the role of strategic partner (α = 0.957).

Some subvariables were declared by Gilbert and Charpentier (2004) as relevant. A subvariable can be measured through several items. Therefore, based on the Ulrich model, 37 items were used to evaluate the HRF performance only.

In this context, global, we used dichotomous and Likert scales according to the subvariable to measure. For instance, existing documentation processes were determined by a dichotomous scale ("Are the following HR Process documented?" Answer: yes / no); a global scale (check the appropriate(s) answer(s) from a list of proposals) was used to determine possible modes of communication between employees and senior managers; while Likert scale was used for other variables.

8 Analysis tools

For data analysis, we used SPSS data processing software. Having defined and coded variables, shouted the data matrix, and determined the appropriate tests... we have conducted various treatments (consistent with the encoding used), namely flat sorting, the cross tabulations, ANOVA analysis (ANalysis of Variance), chi-square tests.... The choice of mathematical tests used is justified by the literature of data analysis within social sciences (Rafael Costa & Masuy Stroobant, 2013).

9 Finding

9.1 Qualitative inquiry: Content analysis results

Regarding HRF structuring and company's performance, three main themes have emerged. Table1 provides a simplified description of the themes that we will discuss in detail later.

Table 1: Principal themes

	Description	Comment example
Theme 1	HRF structuration	*"I cannot speak of an HR function"*
Theme 2	HRF Performance	*" It should be specified department just for HRM management"*
Theme 3	Management practices	"There is no Occupational Medicine System"

9.1.1 Theme 1: HRF structuring

Through the interviews conducted, several respondents noted the lack of structuring at the HR function. As one respondent has mentioned:

"In fact, I cannot speak about an HR function, because as a HR manager I am just managing staff in its preliminary sense. In other words, my job is to check employees' presence, manage payroll and applications having relationship with the staff, display ..."

In addition to that, there was a feeling that the awareness concerning the importance of HR function contribution on the overall performance of the company was generally rare and often insufficient. As a participant pointed out, " It's rarely that we are interested in assessing the contribution of the HR function in the organization's performance "However, the commitment of the high direction to provide a well-structured HR function can be considered as a standard in the other companies that have a long history in integrating their human resources in order to achieve their objectives.

9.1.2 Theme 2: HRF performance

The second theme that emerged is linked to the performance of HRF. In this sense, we observed that it's hard for some companies to apply their internal administrative procedures as a participant bearing witness: *"Administrative procedures often cause cumulated delays"*, or as another also specified: *"even if the procedures are generally clear and well structured, but they are often rigid while being applied...".* Rigidity regarding administrative procedures application is often found in companies with rigid hierarchical cultures.

Documentation is also a weakness that affects HRF performance. In this context, one interviewee noted, *"All the documentation is done by writing to registers or reports, there is no data record using computer tools"* Even in companies where there was a sufficient number of computers available, there was no guarantee that they are used for audio or video content delivery that can be used as traceability tool, or in order to develop management process that is used.

9.1.3 Theme 3: Management practices

The third theme focuses on emerging weaknesses related to management practices. To find out if industrial companies opt for the use of more advanced management practices, we sought how participant companies are applying the instructions of strategic workforce planning (SWP) and those of labor medicine.

This choice was based on the fact that the application of the SWP gives an idea about company's performance regarding administrative issues, while the adoption of labor medicine allows evaluating the enterprise regarding applying and executing the instructions.

Talking about the SWP, an HR manager of a medium sized company said: *"Based on the fact that our company has not yet exceeded the personnel management stage, so we do not apply strategic workforce planning because it requires the prior existence of certain conditions regarding the structuring o f HR function, management tools ...and that's the thing we do not have until now* ". The testimony of this participant emphasized the existence of conditions while applying management methods.

In connection with the occupational medicine which is considered as a maturity indicator in terms of HRM, one respondent said: *"Our occupational medicine system is manifested through a doctor who comes to visit the corporate site every Wednesday. So if an employee has a health problem he can consult the doctor who makes decisions regarding health care, and who decides to allow him to take some rest, or to recommend that the employee changes his job inside the firm if his health case so requires".* Another person reported *"We do not have metrological measures of risks that an employee may encounter within the company".* Similar comments were frequently found particularly among small and medium sized enterprises.

9.2 Finding through the quantitative study

While assessing the HRF, we were inspired by the model of Ulrich (1996) in which he defines four roles of HRF. In this measurement, the results obtained are as follows:

9.2.1 The administrative expert role

By choosing "documentation" as one of the variables measuring the administrative role of the HRF, we noticed that companies are interested first in documenting recruitment processes and remuneration (fixed and variable), they focus then on employees' internal mobility, on documenting staff departures, and trainings process, followed by those related to social relations. Finally, they are interested in career management.

In line with this, 80 interviewees (73.8%) have affirmed that the companies to which they belong have a clear labor regulation that is well respected at an average of 80% by employees.

On the other hand, 78 (71.6%) companies use absenteeism indicators, seniority indicators (66.1%), effective indicators (64.2%), attendance (42.2%) and performance (23.9%) indicators. They also use those related to wages assessment (6.4%), to turn-over rate (6.4%), and revealing outsourced function (7.3%), while only 3.7% establish an index estimating annual number of applications they receive.

Thus, analysis has shown a strong correlation between HRF performance and the "administrative expert role" (r = 0.941; p = 0.000). Figure 1 shows in advantage that the more a company is large-sized, the more the mastery of the administrative aspects promotes the HRF performance.

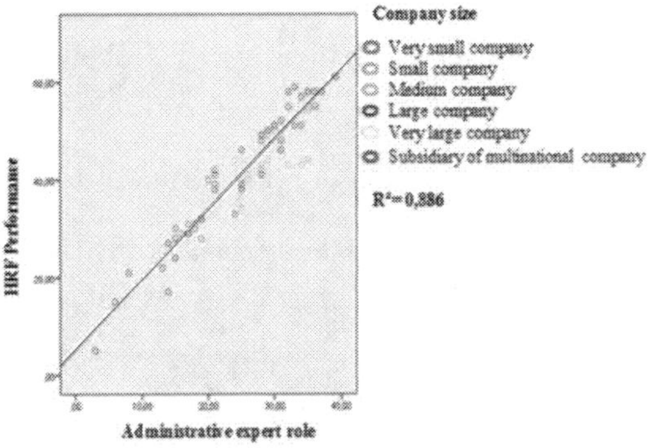

Figure 1. Relationship between HRF performance and the administrative expert role.

9.2.2 The employees' champion role

Among the participating firms in the survey, 18 companies (16.5%) encourage their employees to pursue their academic or professional studies, 59.6% have an integration system for new recruits, while 53.2% favor this integration through implementing a sharing experience system that promotes exchange between former employees and new recruits.

We also found that 75 (68.6%) companies encourage employees to propose solutions to labor problems, 66 (60.6%) enterprises supervise also new ideas' implementation, but only 42.2% (46) ensure the improvement of these solutions.

In order to have more detailed information concerning the contribution of "employees' champion role" in the overall performance of the HRF, we conducted a correlation analysis that results in: r = -0.72;p = 0.472 (non-significant relationship). This shows that the "employees' champion" role cannot be judged as being taken into consideration while promoting the HRF within industrial companies in Morocco. The relationship between "employees' champion role" and HRF performance is influenced by the business type.

9.2.3 The change agent role

As part of evaluating "communication" as being a determinant of change agent (Gilbert and Charpentier, 2004), we found that several communication modes are simultaneously possible. Indeed, 89 (81.7%) managers reported that their collaborators can communicate with them in person (face to face) easily, while 63.6% stated that they can do it by phone. However, within 35 (32.1%) companies, an employee must make a request in order to communicate with the top manager, while in other firms (27.5%), the employee must go through his superior before seeing the top manager.

A correlation test shows the existence of a moderately strong relationship between the performance of HRF and the "change agent role" (r = 0.450), but it is highly significant (p=0.00).

9.2.4 The strategic partner role

"Strategic partner role" has a strong and positive correlation with the HRF performance (r = 0.745; p<0.01). Among seventy two companies (66.1%) adopting a HRF, 52 respondents (47.7%) stated that the HRD takes part in the Management Committee, and participates in strategic decision-making at various frequencies (8.3% sometimes do, 44% are often involved in decision making, 3.7% do so very often).

In terms of control and efficiency of management systems, our survey reveals that only 31.19% of managers are often informed by policy and overall directions of HRM within their companies (Figure 2).

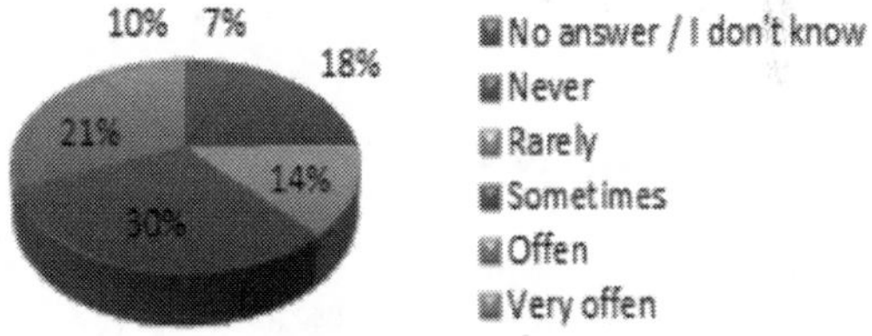

((21%) often and (10%) very often).

Figure 2. Being informed by human resource management policies

Besides; statistics indicate that 40% of participants no longer have the ability to access the information given by the management indicators. Others (18%) benefit from limited information.

The enquiry regarding the effectiveness of certain HRM systems showed that only 30.15% of managers can be judged satisfied regarding management systems within their businesses.

10 Discussion

The relationship between company size and structure of its HR function was predictable. The more a company is large the more we talk about steering committees, and functional departments. Thus a large number of people working in the same department requires some structuring to ensure the smooth progress of work.

However, it was clear that we must not confuse between company size and the adoption of an HR function. Indeed, according to our survey, we found that some small businesses adopt a "HR function", while the others are satisfied with the "personnel management". A slightly more detailed analysis showed that the first half consist of small companies (fewer employees), but are at the same time subsidiaries of multinational companies. This explains the existence of an HR function (a subsidiary of a multinational company) but not well structured (because of the reduced number of employees).

During our investigation, we found also that many respondents were unable to identify whether the HR department adopts a SWP although some spent 10 years in the business. This leads us to question the opening of HR departments to executives and managers of the company, as well as their integration into decision-making.

Based on the finding, it's clear that the HRF is giving more importance to the administrative issues than to the managerial ones. The role of "employees' champion" as well remains limited due to the influence of corporate culture which is closed in most cases.

In such context, many firms will continue to face obstacles because they are willing to be opened up economically but still managing their human resources according to outdate practices.

In this case, the solution remains in seeking the harmonization among the human resources in order to lead the supply chain to perform effectively because researches proved that *"No dominance among supply chain members leads to the highest profit for the entire supply chain"* [8], [9].

11 Conclusion

In conclusion, through this study, we wanted to focus on HRM applied within Moroccan industrial companies and perceived from the point of view of the top managers. Specifically, we have not only referred the assessment of the HR function in terms of structure and maturity, but we also targeted the description of the established management styles. Analyses of results found overturned or confirmed several assumptions.

Thus, involved in achieving the fundamental objectives of internal and external coordination of physical and information flows, responsiveness to hazards, primate of logistical requirements, empirical knowledge of the status of human tissue sites, and adoption of adequate means; the HRM strongly contributes to the determination of practices within the supply chain, both operationally and strategically.

References

[1] Haut-Commissariat au Plan- Royaume du Maroc, *Les sources de la croissance économique au Maroc*, http://www.hcp.ma/file/111438/.

[2] Mentzer, J. T., DeWitt, W., Keebler, J. S., Min, S., Nix, N. W., Smith, C. D., and Zacharia, Z. G. (2001). Defining supply chain management. *Journal of Business logistics*, 22(2), 1-25, 2001.

[3] AFNOR, NF X 50-600, *Logistique – fonction et démarche logistiques*, 2005

[4] Chen, J., Liang, L., Yao, D. Q., & Sun, S. Price and quality decisions in dual-channel supply chains. *European Journal of Operational Research*, 259(3), 935-948, 2017.

[5] Wu, X., and Zhou, Y. The optimal reverse channel choice under supply chain competition. *European Journal of Operational Research*, 259(1), 63-66, 2017.

[6] McAFEE, R. B., Glassman, M., & Honeycutt Jr, E. D. Influence de la culture d'entreprise et des politiques de ressources humaines sur la stratégie Supply Chain Management (SCM). *Logistique & Management*, 11(1), 53-65, 2003.

[7] Choi, E. K., & Joung, H. W. Employee job satisfaction and customer-oriented behavior: A study of frontline employees in the foodservice industry. *Journal of Human Resources in Hospitality & Tourism*, 16(3), 235-251, 2017.

[8] Luo, Z., Chen, X., Chen, J., & Wang, X. Optimal pricing policies for differentiated brands under different supply chain power structures. *European Journal of Operational Research*, 259(2), 437-451, 2017.

[9] Yu, D. Z., Cheong, T., & Sun, D. (2017). Impact of supply chain power and drop-shipping on a manufacturer's optimal distribution channel strategy. *European Journal of Operational Research*, 259(2), 554-563, 2017.

AUTHOR INDEX

EDP Sciences – Web of Conferences
17, Avenue du Hoggar
Parc d'Activité de Courtabouf
BP 112
F-91944 Les Ulis Cedex A
France

ISBN 978-1-5108-3939-7

CURRAN ASSOCIATES INC.
proceedings.com

4th International Symposium on Mechanics, Structures and Materials Science (MSMS 2023)

MATEC Web of Conferences Volume 380 (2023)

Tianjin, China
30 - 31 March 2023

edp sciences